Concerns and Advances of Agricultural Water Management

Concerns and Advances of Agricultural Water Management

Edited by **Keith Wheatley**

New York

Published by Callisto Reference,
106 Park Avenue, Suite 200,
New York, NY 10016, USA
www.callistoreference.com

Concerns and Advances of Agricultural Water Management
Edited by Keith Wheatley

International Standard Book Number: 978-1-63239-127-8 (Hardback)

Printed in the United States of America.

Contents

Preface

Every book is a source of knowledge and this one is no exception. The idea that led to the conceptualization of this book was the fact that the world is advancing rapidly; which makes it crucial to document the progress in every field. I am aware that a lot of data is already available, yet, there is a lot more to learn. Hence, I accepted the responsibility of editing this book and contributing my knowledge to the community.

This book highlights the concerns related to food security and agricultural water management. Food security came up as a problem in the first decade of the 21st century, questioning the sustainability of humankind, which is certainly associated directly to the agricultural water management that has varied dimensions and needs integrative expertise in order to be dealt with. The aim of this book is to integrate the subject matter that deals with sustainable irrigation management & development and strategies for irrigation water supply & conservation in a single text. It is a comprehensive compilation of information regarding content revealing situations from distinct continents. Several case studies have been elucidated in this book to provide the readers with a general scenario of the problem, challenges and perspective of irrigation water use. The book serves as a descriptive reference for professionals, students and researchers working on distinct aspects of agricultural water management.

While editing this book, I had multiple visions for it. Then I finally narrowed down to make every chapter a sole standing text explaining a particular topic, so that they can be used independently. However, the umbrella subject sinews them into a common theme. This makes the book a unique platform of knowledge.

I would like to give the major credit of this book to the experts from every corner of the world, who took the time to share their expertise with us. Also, I owe the completion of this book to the never-ending support of my family, who supported me throughout the project.

Editor

Part 1

Sustainable Irrigation Development and Management

Soil, Water and Crop Management for Agricultural Profitability and Natural Resources Protection in Salt-Threatened Irrigated Lands

Fernando Visconti[1,2] and José Miguel de Paz[2]
[1]*Desertification Research Centre – CIDE (CSIC, UVEG, GV), Valencia*
[2]*Valencian Institute of Agricultural Research - IVIA*
Center for the Development of the Sustainable Agriculture - CDAS, Valencia
Spain

1. Introduction

In the world areas under arid, semi-arid or dry subhumid climate, i.e. where potential evapotranspiration (ETp) exceeds rainfall (R), water scarcity imposes limits on agricultural diversity and productivity. Nevertheless, soils of high potential productivity are also often found under such climates, usually associated to river lowlands where fresh water proximity has allowed irrigation development to produce crops of high nutritional and economic value. It has been estimated that one sixth of world cultivated area is irrigated (AQUASTAT, 2008). What is more important, one third of world agricultural production comes from irrigated lands, and this fraction is going to significantly increase in the upcoming years (Winpenny, 2003). The main restriction to meet all of the soil productive potential of areas where ETp exceeds R is, in addition to water scarcity, soil salinity.

Most of the water nowadays used for irrigation has first originated in rainfall (Fig. 1). The precipitation water on the continents can either infiltrate or run across the rocks and/or soil until it reaches a water body. The infiltrating water into the soils constitutes the soil moisture. It can percolate away from the rooting depth and eventually becomes groundwater. Throughout the soil and ground rocks, water reacts with minerals and as a consequence dissolves salts. Groundwater contributes a significant part of surface water and then, it adds the salts originated in soils and ground rocks. If groundwater does not spring, it continues its movement through the underground rocks usually increasing its load of salts. The salinization of the groundwater occurs due to a lengthy contact with ground minerals, and also because of other phenomena such as contact with saline strata, and seawater intrusion in coastal aquifers. Quite the opposite, the load of salts of surface waters is diluted by direct surface runoff. As a consequence, groundwaters are, in general, more saline than stream waters (Turekian, 1977). Whichever the case, when waters are applied to soils for irrigation, the salts in solution are also applied. Crops absorb water and exclude the major portion of salts, which are left behind in the soil. The absorbed water is transpired to the atmosphere and therefore salts concentrate in the soil solution. Nevertheless, when part of the irrigation water percolates through the bottom of the rooting depth, the salt build-up in soils does not increase indefinitely, it reaches an equilibrium point. This equilibrium point features a steady state, in which the mass of salts entering the soil equals the mass of salts leaving it. This equilibrium point is characterized by a constant medium-to-long-term-average soil salt content.

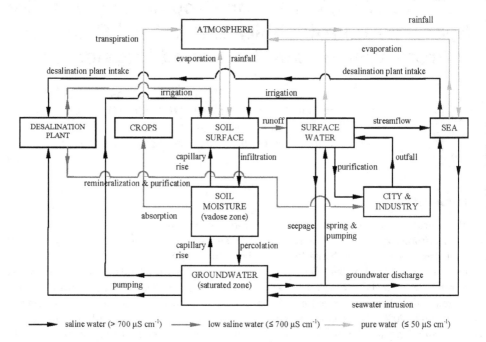

Fig. 1. Agrohydrological cycle

In arid, semi-arid and dry subhumid areas evapotranspiration exceeds precipitation and little water from rainfall percolates through the rooting depth. The more arid is climate the higher is the soil salinity featuring the equilibrium point. The excess of salts is defined with regards to plant tolerance. Plants absorb water from the soil solution, and therefore they respond to the salinity of the soil solution, rather than to the overall salinity of the soil. The salts dissolved in the soil solution decrease the potential of the soil water, which leads to a drought-like situation for plants. Given one plant species, as the soil solution salinity overcomes a plant-characteristic limit the crop suffers from drought and therefore yields decline. A good management of irrigation in arid to dry subhumid areas must provide the plants not only with the water they need to match the crop evapotranspiration, usually called the crop water requirement, but also with some excess water. This extra amount of water leaches, —in arid areas—, or helps to leach, —in semi-arid and dry subhumid areas—, part of the salts carried by the irrigation water itself. In addition to excess irrigation a good drainage must be assured to dispose of the percolating water. This way drainage complements irrigation to achieve a sustainable irrigation management.

The salinity of water systems including soil solution is made up mainly of only eight inorganic ions: sodium (Na^+), chloride (Cl^-), calcium (Ca^{2+}), magnesium (Mg^{2+}), sulphate (SO_4^{2-}), bicarbonate (HCO_3^-), potassium (K^+), and often also nitrate (NO_3^-). As charge bearing

Soil, Water and Crop Management for Agricultural Profitability and Natural Resources Protection in Salt-Threatened Irrigated Lands

5

particles these ions give the water where they are dissolved the property to conduct electricity. Therefore the electrical conductivity at 25° C (EC_{25}), usually in units of dS m^{-1} or μS cm^{-1}, is commonly used as a measure of the salinity of water systems including soil solutions and irrigation waters. The ions just indicated combine to form several salts that differ in their solubility from the low to moderate solubility of calcite ($CaCO_3$) and gypsum ($CaSO_4 \cdot 2H_2O$) to the high solubility of the sodium and chloride salts. Precipitation of calcite and gypsum prevents the salinity of the soil solution from attaining harmful values when calcium, bicarbonate and / or sulphate are concentrated enough in the irrigation water. In addition to this favourable effect on salinity, calcite and gypsum have also a favourable effect on the soil cation balance. The combination of low salinity with a relatively high concentration of sodium with respect to calcium and magnesium, which is traditionally accounted for by the sodium adsorption ratio (SAR = $[Na^+]/([Mg^{2+}] + [Ca^{2+}])^{1/2}$), harms soil structure with consequences on water infiltration and soil aeration. High SAR values have also harmful effects on plants independently of salinity, because of the nutritional imbalance caused by the excessive concentration of sodium with regard to calcium. The weathering of calcite and/or gypsum from soil materials increases the calcium and sometimes magnesium content of the soil solution counteracting, on the one hand, the damage low salinity and high SAR have on soil structure, and on the other hand, counteracting the damage caused on the plant by a sodium high soil solution.

According to the sensitivity analysis of the steady-state soil salinity model SALTIRSOIL the expected average soil solution salinity depends on three main factors: climate, irrigation water salinity and irrigation water amount in this order (Visconti et al., 2011a). Traditionally, farmers have acted on these three factors to gain control on soil solution salinity.

Control over precipitation is out of human reach, however, farmers have some control on soil's climate. All the water saving practices aimed at increasing water infiltration and decreasing water evaporation help decrease also soil salinity (Zribi et al., 2011). Soil infiltration is traditionally enhanced by tillage and mulching with coarse materials of organic and inorganic origin. Soil evaporation is diminished through suppression of weed growth, irrigating at night and mulching with the same materials as before in addition to plastic mulches.

Regarding water quality, farmers have little control on the salinity of a given water body. Surface water has been traditionally the first and usually only option for irrigation. However, other water supplies have been made available throughout history thanks to collective initiatives led by irrigators unions, governments and enterprises. Rainwater harvesting (Huang et al., 1997; Abdelkhaleq & Ahmed, 2007) and water diversions have been used in many instances as non-conventional water supplies well before the 20th century. Groundwater has been used for millennia to irrigate where surface water was absent. However, the intensive exploitation of groundwater resources for irrigation did not occur until the late 19th century when the powerful machinery necessary for drilling and pumping water from depths beneath 8 m was available (Narasimhan, 2009). Other non-conventional water resources have arisen during the 20th century such as waste and reclaimed waters of urban, industrial and mining origin and also desalinated waters. Each one of these water supplies is characterised by a different composition and therefore salinity and SAR. Traditionally farmers have not been aware of these differences until the effects on plants have revealed themselves. Nowadays measurement of, at least, surface water salinity is often routinely carried out by government authorities and irrigators unions. Although farmers cannot change the quality of a water body, modern irrigation methods have allowed them changing the quality of the water actually used for irrigation. This is usually done by fertigation, but also by blending waters from different sources in irrigation reservoirs. The same technology available for fertigation can be used for adding chemicals such as gypsum or mineral acids to decrease the soil solution SAR if necessary.

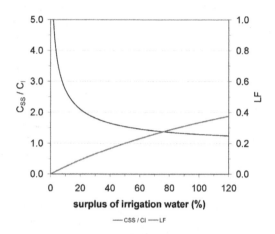

Fig. 2. Leaching fraction (LF) and relative salinity of the soil solution (C_{SS} / C_I) as function of the surplus of irrigation water following a 40:30:20:10 root water uptake pattern and a quotient R / ET_c of 0.5. Equations after Hoffman & Van Genuchten (1983)

The irrigation water amount is not as influential on soil salinity as climate and irrigation water quality. However, this factor has been traditionally considered as the one through which the farmer can exert more control over soil salinity. The idea that irrigation water leaches soil salts has established itself in many places as the popular belief that the more you irrigate the more salts you leach out of the soil. However, the relationship between soil salinity and irrigation water amount is far from being linear. The relationship is in fact a rational in which once the sum of rainfall and irrigation have matched the crop water requirement the soil solution salinity rapidly decreases with irrigation water surpluses of only 10 to 20% (Fig. 2). From 30% on, the soil solution salinity hardly decreases. It tends asymptotically to a limit which depends on climate, specifically the quotient rainfall to evapotranspiration, and irrigation water salinity.

Not only excess overirrigation constitutes a waste of water, which is on itself a severe problem in the present global scenario of scarcity and competition for safe water resources. As occurs with overfertilization it can be self-defeating. The amount of irrigation water must not surpass the limits imposed not only by the availability of water resources, but also by the capability of the drainage systems and the hydrology of the whole area where the crops are grown. In the medium to long term overriding the natural and man-made irrigation and drainage limits gives rise to serious on and off-farm problems of degradation of lands and water bodies (rivers, lakes and aquifers). Among these problems caused by overirrigation we find the rise of the water table underlying the crop fields, which impedes the soil leaching and leads to waterlogging and soil salinization. Furthermore, overirrigation increases the amount of drainage effluents, which are usually loaded with salts, nutrients and agrochemicals. This constitutes on the one hand a waste of farm investment, and on the other, a potential damage to the natural water bodies because of salinization, eutrophication and pollution.

Provided excess overirrigation is far from being adequate either in terms of agricultural profitability or natural resources protection, the question is how much water in excess of the crop water requirement is necessary to keep soil salts below the limit from which yields will decline. This question has been traditionally answered performing the following calculation (Eq. 1), where IR, R and ET_c are for the irrigation requirement, rainfall and crop evapotranspiration respectively, all in units of L T^{-1}, usually mm yr^{-1}.

$$IR = \frac{ET_c}{1 - LR} - R \tag{1}$$

Providing the amount of water that percolates through the bottom of the root zone is the soil drainage (D), the fraction of the infiltrating water ($I + R$) which becomes the soil drainage is known as the leaching fraction ($LF = D / (I + R)$) where I is for the actual irrigation. In Eq. 1 LR stands for the leaching requirement, which is defined as the minimum leaching fraction necessary to leach the soil salts below a limit considered harmful for a given crop. In order to optimize the irrigation rates the calculation of the leaching requirement has been the objective of several simulation models during the last 50 years.

The irrigation scheduling based on the calculation of a leaching requirement assume at least that i) the steady-state hypothesis is valid enough for the irrigation project, and ii) that the farmer has enough control over the irrigation application to adjust the quantity of water delivered to the soil. The steady-state hypothesis has been criticized because soil salinity fluctuates heavily in the short term following mostly the soil water content. Nevertheless, the leaching requirement is not intended to be a parameter useful at little scale either in time or spatial terms. Rather the leaching requirement is useful for irrigation planning from months to years, and from plots to irrigation districts. Ideally, how the irrigation rates and scheduling should be applied would start from the knowledge of the maximum soil salinity tolerable by the crop or crops to be cultivated during the whole growing season. Next the annual leaching requirement would be assessed with a model such as the traditional LR model (Rhoades, 1974), the WATSUIT (Rhoades & Merrill, 1976) or another developed for the same purpose. Accurate enough predictions of soil salinity only demand i) annual averaged boundary conditions, ii) a coarse spatial discretization, and the simulation of iii) cation exchange and iv) gypsum dissolution–precipitation (Schoups et al., 2006). WATSUIT has the characteristics (i) and (ii) and simulates gypsum equilibrium chemistry. Therefore, despite the last version of WATSUIT is 20 years old, it continues to be a benchmark for developing irrigation guidelines for salt-threatened soils. Once the leaching requirement is known, the required amount of irrigation water can be calculated by means of Eq. 1. Nevertheless, as weather varies from year to year how this amount of water has to be applied demands knowledge about soil water content. This knowledge can be based on meteorological data and soil water content measurements. All these in addition to farmers' experience should guide the application of irrigation water.

The model SALTIRSOIL was originally developed for the simulation of the annual average soil salinity in irrigated well-drained lands (Visconti et al., 2011b). It has characteristics similar to WATSUIT. The input data to the model included i) climate data such as monthly values of reference evapotranspiration (ET_0) and amount and number of days of rainfall, ii) water quality data such as yearly average concentrations of the main ions, iii) irrigation scheduling data such as monthly values of irrigation amount, number of irrigation days and percentage of wetted soil, iv) crop data such as monthly or season basal crop coefficients, percentage of canopy ground cover and sowing and harvest dates for annual crops, and finally v) chemical and hydrophysical soil data. The SALTIRSOIL was intended to be a predictive model, however, it can be used for irrigation and soil management. The best irrigation scheduling for keeping soil salinity below some critical value can be found batch running the same simulation while changing the irrigation rates and schedule.

Following the methodology just described the SALTIRSOIL model is useful to search for the most adequate irrigation rates and scheduling in order not to surpass an average-annual limit of soil salinity. This is interesting but it could be improved without any loss of the original applicability of the model, i.e. optimum ratio of information to data requirements. This has been done adapting the SALTIRSOIL algorithms for the monthly average calculation of soil salinity.

In the following the new algorithms implemented in SALTIRSOIL for the calculation of the monthly average soil salinity in irrigated well-drained lands, and the use of this new SALTIRSOIL, from now on referred to as the SALTIRSOIL_M model, for the development of optimum guidelines for soil, water and crop management in irrigated salt-threatened areas will be shown. These guidelines will be discussed in the framework of the different productive and environmental challenges irrigation faces in a relevant place in SE Spain.

2. SALTIRSOIL_M: A new tool to assess monthly soil salinity and for irrigation management in salt-threatened soils

The SALTIRSOIL was developed as a deterministic, process-based and capacity-type model. The development of the SALTIRSOIL model started from the characteristics that made the steady-state models WATSUIT (Rhoades & Merrill, 1976) and that of Ayers & Westcot (1985) so useful for the leaching requirement calculation and for assessing the water quality for irrigation.

Steady-state models for soil salinity start from the hypothesis that soil water and salt content keep constant through time. These conditions could only be true if water would continuously flow through soil. This is never the case because irrigation and rainfall are discontinuous processes. Modern transient-state models take into account the time variable, which makes them able to give accurate values of soil water and salt content as has been shown by Goncalves et al. (2006) for the HYDRUS model. Despite these advantages, transient-state models are seldom used outside of research applications because they demand data not available or difficult to obtain. The time variable can be, however, implemented in soil salinity steady-state models while preserving their basic assumptions. This has been shown by Tanji & Kielen (2002), and on a daily basis by Isidoro & Grattan (2011).

The original SALTIRSOIL model has been adapted for the monthly calculation of soil salinity to give the SALTIRSOIL_M model. Therefore the new SALTIRSOIL_M performs a water and salt balance in monthly steps. In the simulations the soil is divided in a number of layers selected by the user. In each simulation the water balance is calculated first, and then

Soil, Water and Crop Management for Agricultural Profitability and Natural Resources Protection in Salt-Threatened Irrigated Lands

9

the soil solution concentration factor of the soil solution regarding the irrigation water in each layer. An average soil solution concentration factor for each month is calculated afterwards. The composition of the irrigation water each month is multiplied by the corresponding monthly average concentration factor and the calculation of the composition of the soil solution at different soil water contents and allowing to equilibrate with soil CO_2, calcite and gypsum is carried out. Finally the electrical conductivity at 25 °C is assessed. The SALTIRSOIL model concepts for the annual calculation of the soil salinity have been described in detail elsewhere (Visconti et al., 2011b). Here only the calculations implemented in SALTIRSOIL_M for the monthly balance of salts in the soil solution are shown.

2.1 Monthly mass balance of salts in the soil solution

Let the soil be split in a number n of layers, and let the shallowest soil layer be the layer 1. The mass of a conservative solute in the solution of the layer 1 in the month i ($m_{i,1}$) can be calculated from Eq. 2.

$$m_{i,1} = m_{i-1,1} + I_i C_{Ii} - D_{i,1} C_{i,1} \qquad (2)$$

Where $m_{i-1,1}$ is the mass of the solute in layer 1 the previous month ($i - 1$), I_i and $C_{i,1}$ are, respectively, the amount of irrigation water and the concentration of the conservative solute the month i, and $D_{i,1}$ and $C_{i,1}$ are the drainage from the layer 1, and the concentration of the solute in the soil water in that layer.

The concentration of the conservative solute in the soil solution of the layer 1 is obtained through Eq. 3 where the mass of the solute given by Eq. 2 has been divided by the average water content of that layer the month i ($V_{i,1}$).

$$C_{i,1} = C'_{i-1,1} + \frac{I_i C_{Ii}}{V_{i,1}} - \frac{D_{i,1} C_{i,1}}{V_{i,1}} \qquad (3)$$

Equation 3 can be reorganized to isolate the concentration of the solute as a function of the rest of variables (Eq. 4).

$$C_{i,1} = \frac{C'_{i-1,1} V_{i,1} + I_i C_{Ii}}{V_{i,1} + D_i} \qquad (4)$$

In Eq. 3 and Eq. 4 $C'_{i-1,1}$ is the mass of solute the previous month divided by the volume of soil water in that layer the present month i. This variable can be expressed in terms of the concentration of the solute in the layer 1 the previous month considering the quotient of the soil water the previous month and the present month (Eq. 5).

$$C'_{i-1,1} = C_{i-1,1} \frac{V_{i-1,1}}{V_{i,1}} \qquad (5)$$

Eq. 5 is substituted in Eq. 4 and after dividing by C_{Ii} Eq. 6 is obtained for the calculation of the concentration factor of the soil solution in layer 1 the month i at average field water content ($f_{i,1} = C_{i,1} / C_{Ii}$).

$$f_{i,1} = \frac{f_{i-1,1}V_{i-1,1}\dfrac{C_{li-1}}{C_{li}} + I_i}{V_{i,1} + D_{i,1}} \tag{6}$$

Similarly to Eq. 2 the mass of a conservative solute in the soil water of a layer j ($j \neq 1$) is calculated with the following equation (Eq. 7).

$$m_{i,j} = m_{i-1,j} + D_{i,j-1}C_{i,j-1} - D_{i,j}C_{i,j} \tag{7}$$

Where $D_{i,j}$ and $D_{i,j\text{-}1}$ are respectively the drainage water the present month i from the layer j and from its overlying layer ($j - 1$), and $C_{i,j}$ and $C_{i,j\text{-}1}$ are the solute concentration the present month i in the layer j and in its overlying layer $j - 1$. Following similar steps to those heading to Eq. 6 we get to Eq. 8 for the calculation of the concentration factor of a conservative solute in the soil water of a layer j in the month i.

$$f_{i,j} = \frac{V_{i-1,j}f_{i-1,j}\dfrac{C_{li-1}}{C_{li}} + D_{i,j-1}f_{i,j-1}}{V_{i,j} + D_{i,j}} \tag{8}$$

2.2 Development of irrigation recommendations: A case study for several crops in the traditional irrigated area of *Vega Baja del Segura* (SE Spain)

The SALTIRSOIL_M model has been used to develop irrigation recommendations in the relevant traditional irrigated district of *Vega Baja del Segura* (SE Spain).

The *Segura* River and *Baix Vinalopó* lowlands together represent one of the most important agricultural areas in Spain. More than 90% of the land is irrigated and approximately 80% of it is salt-affected (de Paz et al., 2011). The main crops that cover 61% of the irrigated area are citrus such as orange, mandarin and Verna lemon (*Citrus sinensis, Citrus reticulata* and *Citrus limon* (L) Burm f.) grafted onto various different rootstocks. The moderately salt-tolerant Sour Orange (*Citrus aurantium* L.) and especially Cleopatra mandarin (*Citrus reshni* Hort. ex *Tan.*) are used as rootstocks for more than 60% of citrus. Vegetables (including tubers) cover 16% of the area. These are globe artichoke (*Cynara scolymus* L.), lettuce (*Lactuca sativa* L.), melon (*Cucumis mello* L.), broccoli (*Brassica oleracea*, Botrytis group), and potato (*Solanum tuberosum* L.). Non-citrus fruit trees cover 12% of the area, specifically almond (*Prunus dulcis*), pomegranate (*Punica granatum* L.) and date palm (*Phoenix dactylifera* L.).

The *Segura* River and *Baix Vinalopó* lowlands comprise several irrigation districts, each one of them featured by different irrigation systems, crops and water supplies. The traditional irrigation district of *Vega Baja del Segura* (Fig. 3) is one of the most important because of the use of water resources, which has been estimated between 80 and 120 hm³ yr⁻¹ (Ramos, 2000), number of farmers, productivity, history and the large stretch of land, which amounts up to approximately 20000 ha from which 15000 ha are actually irrigated each year (MMA, 1997). The average Penman-Monteith reference evapotranspiration and precipitation are 1215 and 385 mm yr⁻¹, respectively. In this irrigation district the distribution of horticultural and tree crops is 70-30% (MMA, 1997). The main irrigation water supply in the irrigation district is the *Segura* River itself. Although new irrigation projects use drip systems, at least 50% of the area is still irrigated by surface.

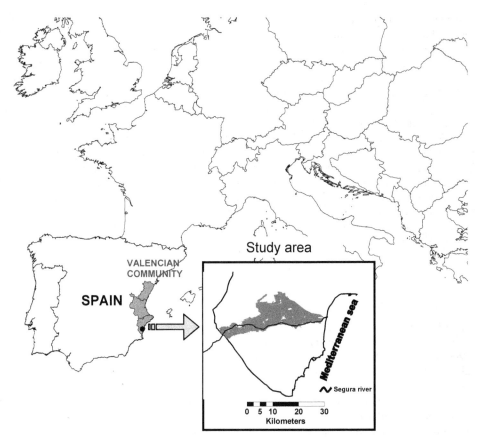

Fig. 3. Location of the Traditional Irrigation Area of the *Vega Baja del Segura*

Month	pH	Alk.	Na⁺	K⁺	Ca²⁺	Mg²⁺	Cl⁻	NO₃⁻	SO₄²⁻	EC₂₅
Jan	7.75	5.75	20.84	0.52	6.24	5.87	17.22	0.59	9.62	3.91
Feb	7.77	5.85	21.84	0.50	7.22	6.93	19.53	0.52	11.33	4.19
Mar	8.10	5.42	20.76	0.57	7.34	7.09	21.81	0.66	12.41	4.41
Apr	7.85	5.81	24.67	0.63	7.49	7.47	24.14	0.51	12.03	4.92
May	7.80	5.78	22.13	0.45	6.60	6.77	20.27	0.34	9.77	4.49
Jun	7.48	5.03	16.94	0.44	5.80	5.74	16.87	0.22	9.56	3.51
Jul	7.50	5.93	39.71	0.68	7.13	8.21	39.83	0.38	12.42	6.39
Aug	7.63	5.16	30.42	0.48	5.81	6.23	28.10	0.38	10.92	4.89
Sep	7.78	4.69	16.65	0.40	5.03	5.16	15.25	0.24	7.90	3.38
Oct	7.60	5.37	27.15	0.60	6.82	6.68	23.77	0.65	11.52	4.46
Nov	7.88	4.96	19.86	0.52	5.88	5.88	17.49	0.65	10.19	3.87
Dec	7.64	5.02	20.58	0.52	6.54	6.54	18.80	0.64	11.15	4.09
Avg.	7.73	5.40	23.46	0.53	6.49	6.55	21.92	0.48	10.73	4.37

Table 1. Monthly characteristics of the Segura River water during the three year period 2007-2009. All ion concentrations in mmol L⁻¹, EC₂₅ in dS m⁻¹ and alkalinity (Alk.) in mmol_C L⁻¹

The *Segura* River goes through the traditional irrigated district of *Vega Baja del Segura*, and there exhibits annual averages of electrical conductivity at 25 °C (EC$_{25}$) and SAR of 4.3 dS m^{-1} and 6.3 (mmol L^{-1})$^{1/2}$, respectively. However, the EC$_{25}$ and SAR remarkably fluctuate through the year (Table 1) following the cycle of water releases from upstream dams (Ibáñez & Namesny, 1992). From late autumn till mid spring water is slowly released from dams to maintain environmental flow, which includes winter irrigation. Important water releases start in spring, and along with them the EC$_{25}$ slightly increases because the low EC$_{25}$ water ($\approx$ 1.2 dS m^{-1}) of the upstream dams helps sweep the outfalls from the sewage treatment plants and irrigation returns through a river that otherwise presents a constant but low base-flow. The next months, the EC$_{25}$ decreases until it reaches a minimum in June. During July the EC$_{25}$ increases again because the irrigation returns from upstream lands increase the river flow and because water releases stop during this month. In late July and early August the important water releases resume and the EC$_{25}$ decreases again until it reaches another minimum in September. Then the important water releases stop until the next year and the EC$_{25}$ attains a maximum during October because of the autumn rainfalls. This is the most important rainfall season in the area and it effectively leaches the salts from the lands as the increase in the EC$_{25}$ of the river shows. Because of the correlation between electrical conductivity and sodium adsorption ratio in the *Segura* River the SAR follows a parallel fluctuation to the EC$_{25}$.

2.2.1 Set up of simulations

The soil saturation extract composition of the soils of the *Vega Baja del Segura* was simulated with SALTIRSOIL_M under ten different crops. These were three horticultural crops and seven tree crops. The horticultural crops were globe artichoke, grown from October 1st until July 8th, and rotation of melon and broccoli, from September 14th until January 27th, and melon also from April 1st until August 19th and potato from September 14th until January 22nd. The tree crops were date palm, sweet orange, lemon grafted onto sour orange, lemon grafted onto Mandarin Cleopatra, lemon grafted onto *Cytrus Macrophylla*, Verna lemon and pomegranate. These ten crops are representative of at least 75% of the agriculture of the *Vega Baja del Segura* and according to their threshold-slope functions of yield against electrical conductivity of the saturation extract (EC$_{se}$) they exhibit different tolerances to soil salinity (Figure 4). Except for date palm and globe artichoke which are from moderately tolerant to tolerant, the rest of crops are moderately sensitive to soil salinity. Pomegranate is between moderately sensitive to moderately tolerant defining in fact the limit between both categories. The data on soil, climate, threshold-slope functions and basal crop coefficients used in the simulations can be found in Visconti et al. (2012).

Simulated crop	Jan	Feb	Mar	Apr	May	Jun	Jul	Aug	Sep	Oct	Nov	Dec	Tot
Artichoke	1	25	32	70	89	81	31	7	0	0	4	0	339
Mel.-Broccoli	0	0	0	34	65	110	122	68	0	0	8	2	410
Melon-Potato	0	0	0	34	65	109	121	68	0	0	20	6	423
Date palm	2	23	27	73	97	135	139	122	22	0	18	3	660
Sweet orange	0	15	14	50	65	90	98	93	1	0	13	0	440
Lemon trees	0	9	4	39	51	81	84	72	0	0	7	0	347
Pomegranate	0	0	0	44	66	102	106	92	2	0	2	0	415

Table 2. Crop water requirements in mm calculated with SALTIRSOIL

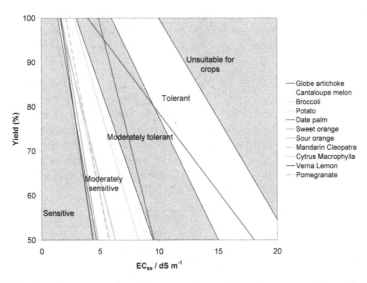

Fig. 4. Threshold-slope functions of yield versus electrical conductivity of the soil saturation extract (EC$_{se}$) for the crops simulated with SALTIRSOIL_M in the *Vega Baja del Segura*. Categories after Maas & Hoffman (1977)

For each one of the crops the annual leaching requirement was assessed in the following way. The crop water requirement, i.e. the crop evapotranspiration, was first calculated with the SALTIRSOIL (Table 2). Then starting with the simulation in which the irrigation dose was set equal to between – 50 to – 30% of the crop water requirement, several simulations were carried out gradually increasing the annual irrigation water amount in 5% steps. Once the batch of simulations was finished the monthly values of EC$_{25}$ in each simulation were averaged to obtain the corresponding annual EC$_{25}$. As the annual LF is also calculated by the SALTIRSOIL_M, this allowed us to have the graph of annual EC$_{se}$ against LF. The electrical conductivity for 90% yield, which is called EC$_{90}$, was then calculated from the corresponding threshold-slope functions (Fig. 4). For the horticultural crop rotations the EC$_{90}$ was calculated for both crops, and the value for the most sensitive was used, i.e. the lower EC$_{90}$. These were melon and potato for the melon-broccoli and melon-potato rotations respectively. The values of EC$_{90}$ (Table 3, second column) were then interpolated in their corresponding graphs of annual EC$_{se}$ against LF to obtain the annual leaching fraction for 90% yield (LF$_{90}$). This value was taken as the leaching requirement, i.e. LR = LF$_{90}$.

2.2.2 Results of the simulations

The leaching requirements calculated with the SALTIRSOIL_M were between 0.08 and more than 0.99 (Table 3, last column). The moderately tolerant to tolerant globe artichoke and date palm presented leaching requirements of 0.08 and 0.09 respectively. The moderately sensitive to tolerant pomegranate presented a leaching requirement of 0.19. The melon-potato, sweet orange, lemon grafted onto sour orange and onto *Cytrus Macrophyla*, and Verna lemon presented values higher than 0.99. This means that a yield of at least 90% can not be achieved for these crops in the area when irrigating with *Segura* River water. With a leaching requirement of 0.75 the only citrus that could be grown for at least 90% yield with *Segura* River water would be those grafted onto the Mandarin Cleopatra rootstock. With a leaching requirement of 0.50 the succession of melon and broccoli could also be grown for at least 90% yield.

The results of the SALTIRSOIL_M model for the annual leaching requirement were compared with previously calculated leaching requirements with the WATSUIT and the SALTIRSOIL models (Visconti et al., 2012). The SALTIRSOIL_M calculates lower leaching requirements than the SALTIRSOIL as is shown in Table 3. The leaching requirements calculated with SALTIRSOIL_M are also lower than the corresponding values calculated with WATSUIT when dealing with the moderately sensitive to tolerant crops. When dealing with moderately sensitive crops the leaching requirements calculated with SALTIRSOIL_M are higher than the values calculated with WATSUIT.

Simulated crop	$EC_{90}/$ dS m^{-1}	WATSUIT	SALTIRSOIL Surface	SALTIRSOIL_M Surface	SALTIRSOIL_M Drip
Globe artichoke	5.83	0.13	0.10	0.08	0.07
Melon-broccoli	3.55	0.42	0.67	0.50	0.47
Melon-potato	2.53	0.79	> 0.99	> 0.99	> 0.99
Date palm	6.80	0.09	0.09	0.09	0.08
Sweet orange	2.33	0.92	> 0.99	> 0.99	> 0.99
Lemon onto SO	2.48	0.82	> 0.99	> 0.99	> 0.99
Lemon onto MC	2.81	0.65	> 0.99	0.75	0.73
Lemon onto CM	1.72	> 0.99	> 0.99	> 0.99	> 0.99
Verna Lemon	2.19	> 0.99	> 0.99	> 0.99	> 0.99
Pomegranate	4.30	0.27	0.25	0.19	0.17

Table 3. Electrical conductivities for 90% yields (EC$_{90}$) and corresponding leaching requirements calculated with the WATSUIT, SALTIRSOIL and SALTIRSOIL_M models

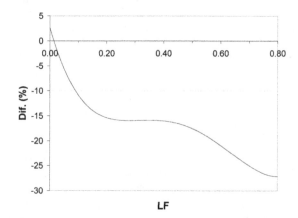

Fig. 5. Average percentage difference (Dif. (%)) between the EC$_{se}$ calculated with SALTIRSOIL and SALTIRSOIL_M as a function of the leaching fraction (LF)

The SALTIRSOIL_M calculates average annual soil salinities between 10 and 30% lower than the SALTIRSOIL as is shown in Fig. 5. Transient-state models calculate lower soil salinities than steady-state models (Corwin et al., 2007). This fact makes the leaching requirements calculated with transient-state models to be lower than the leaching requirements calculated with steady-state models (Corwin et al, 2007; Letey et al., 2011). The implementation of the time variable as simple monthly steps in the SALTIRSOIL_M model has suffice to have soil salinities very similar to those calculated with other more complex and data-demanding transient-state models.

2.2.3 Proposal of irrigation recommendations

The irrigation requirements for the crops for which the 90% yield is achievable are between the 357 mm yr^{-1} of the moderately tolerant artichoke and the 2345 mm yr^{-1} of the moderately sensitive lemon grafted onto the Mandarin Cleopatra rootstock (Table 4). According to the *Segura* Valley Authority (MMA, 1997) and Ramos (2000) the average availability of water for irrigation from the *Segura* River in the traditional irrigated area of *Vega Baja del Segura* can be estimated between 530 and 800 mm yr^{-1}. Assuming that the maximum availability of irrigation water from the *Segura* River is never going to be higher than 800 mm yr^{-1}, the resulting soil salinity (EC_{se}) would be between the EC_{90} of the tolerant date palm (6.8 dS m^{-1}) and the 3.2 dS m^{-1} of the lemon trees (Table 4). The surface weighted average soil salinity would result to be 4.4 dS m^{-1}, with a monthly maximum of 7.3 dS m^{-1} for date palm orchards and a minimum of 2.8 dS m^{-1} for lemon trees orchards. Given this availability of water for irrigation the surface weighted average yield would be 82% with a minimum of 63% for sweet orange orchards. These yields would be achieved with an average 683 mm yr^{-1} of water, i.e., 102 hm^3 yr^{-1} for the whole irrigation district.

In the traditional irrigated area of *Vega Baja del Segura* almost all of the land is equipped with underground pipelines to collect of the waters that percolate through the rooting depth. The drainage waters are disposed by means of a hierarchical system of canals. The major canals are called *azarbes* and they go through the *Vega Baja* more or less parallel to the *Segura* River bed until they pour into the river mouth itself. According to the SALTIRSOIL_M calculations the drainage effluents from the traditional irrigated area of the *Vega Baja del Segura* would be between 61 and 513 mm yr^{-1}, with a surface weighted average of 338 mm yr^{-1}. This would amount to 51 hm^3 yr^{-1} of drainage effluents from the whole district. These drainage effluents would present a salinity (EC_{dw}) between 7 and 24 dS m^{-1}, while the sodicity (SAR_{dw}) would be between 9 and 24 (mmol L^{-1})$^{1/2}$ with weighted averages of 8.3 dS m^{-1} and 10.4 (mmol L^{-1})$^{1/2}$, respectively. These drainage effluents are, thus, high in EC and SAR and become an environmental concern. In spite of their salinity and sodicity, along their way through the district the irrigation returns from upstream lands are usually used again for irrigation (Abadía et al., 1999). Accordingly, on the one hand the district's irrigation water requirement would be less as an important part of the drainage water is reused, and on the other hand, the irrigation application in the moderately tolerant to tolerant crops in the area, i.e. artichoke, date palm and pomegranate, should increase a bit in order to have drainage effluents lower in salts and sodium. It is reasonable to think that both facts would compensate each other and the appropriate irrigation requirement for the whole area should not be less than 102 hm^3 yr^{-1}. Regarding the citrus trees the moderately sensitive sweet and *Cytrus Macrophyla* oranges and Verna lemon should be grafted onto more tolerant rootstocks such as Mandarin Cleopatra, sour orange and other similar to these. With these rootstocks citrus yields of 80-85% would be achievable with just 800 mm yr-1 of Segura River water. These little decrements in citrus yields are usually reflected in decreased average fruit size, however, they are also accompanied by higher juice sugar and acid contents (Grieve et al., 2007). Increments in fruit quality with slight salinity stress have been described for other fruits including melon (Bustan et al., 2005).

The traditional irrigated area of *Vega Baja del Segura* has been irrigated by surface for centuries. Nevertheless, since the early nineties localized irrigation systems are slowly replacing them. Localized irrigation systems are characterized by i) more frequent irrigations, ii) less water application in each irrigation, and iii) less wetted area. The effect of these three variables can be simulated with SALTIRSOIL and SALTIRSOIL_M.

Drip irrigation was simulated in SALTIRSOIL_M decreasing the wetted soil area from 40% to 3% and multiplying the number of irrigation days a year by 6. The irrigation amount was kept constant.

Simulated crop	IR_{90}	I_{rec}	ET_a	D	EC_{se}	$EC_{se}min$	$EC_{se}max$	Y(%)	EC_{dw}	SAR_{dw}
Globe artichoke	357	357	681	61	5.83	5.17	6.64	90	17.7	18.0
Melon-broccoli	1066	800	731	453	3.92	3.37	4.57	87	7.48	9.79
Melon-potato	—	800	748	437	3.97	3.36	4.61	73	7.68	9.93
Date palm	744	744	1032	97	6.80	6.55	7.26	90	23.9	24.3
Sweet orange	—	800	801	383	4.04	3.45	4.60	63	8.84	10.9
Lemon onto SO	—	800	672	513	3.22	2.75	3.82	82	7.27	9.41
Lemon onto MC	2345	800	674	510	3.22	2.75	3.82	84	7.27	9.41
Lemon onto CM	—	800	672	513	3.22	2.75	3.82	69	7.27	9.41
Verna Lemon	—	800	672	513	3.22	2.75	3.82	71	7.27	9.41
Pomegranate	528	528	736	177	4.30	3.55	5.04	90	11.5	13.1
AVERAGES	—	[a]683	[a]729	[a]338	[a]4.40	—	—	[a]82	[b]8.3	[b]10.4

[a]Surface weighted average (70% horticultural, 30% trees), [b]Surface and drainage weighted average

Table 4. Irrigation requirement for 90% yield (IR_{90}), recommended irrigation (I_{rec}), actual evapotranspiration (ET_a), and drainage (D) all in mm yr-1, EC (dS m-1) and SAR ((mmol L-1)1/2) of the saturation extract and of the drainage water calculated for surface irrigation

The leaching requirement for drip irrigation slightly decreases regarding surface irrigation as is shown in Table 3. This occurs because drip irrigation minimizes the evaporation of water from the soil. Therefore, the actual evapotranspiration would drop from 729 to 683 mm yr-1 (Table 4 and Table 5), thus increasing the drainage from 338 to 369 mm yr-1. If the whole irrigation district used drip irrigation systems the irrigation water demand would drop to 667 mm yr-1, i.e., 100 hm3 yr-1. The soil salinity would also drop to 4.3 dS m-1, with a maximum of 7.5 dS m-1 and a minimum of 2.6 dS m-1. Furthermore the yields for citrus would rise and the overall average relative yields would keep or increase. On the other hand the amount of drainage effluents would rise to 55 hm3 yr-1 with average electrical conductivity and sodium adsorption ratio of 8.4 dS m-1 and 10.8 (mmol L-1)1/2, i.e., with salinity and sodicity slightly higher than when using surface irrigation systems.

Simulated crop	IR_{90}	I_{rec}	ET_a	D	EC_{se}	$EC_{se}min$	$EC_{se}max$	Y(%)	EC_{dw}	SAR_{dw}
Globe artichoke	317	317	649	52	5.83	5.09	6.71	90	19.30	19.2
Melon-broccoli	820	800	677	508	3.67	3.03	4.35	89	6.99	9.35
Melon-potato	—	800	694	491	3.72	2.99	4.40	76	7.14	9.46
Date palm	688	688	986	87	6.80	6.47	7.51	90	24.52	24.9
Sweet orange	—	800	730	454	3.70	3.04	4.33	68	7.85	9.73
Lemon onto SO	—	800	632	553	3.12	2.60	3.73	83	6.93	9.12
Lemon onto MC	2024	800	631	554	3.12	2.60	3.73	86	6.93	9.12
Lemon onto CM	—	800	632	553	3.12	2.60	3.73	70	6.93	9.12
Verna Lemon	—	800	632	553	3.12	2.60	3.73	73	6.93	9.12
Pomegranate	435	435	683	136	4.30	3.52	5.29	90	12.21	13.9
AVERAGES	—	[a]667	[a]683	[a]369	[a]4.26	—	—	[a]83	8.4	10.8

[a]Surface weighted average (70% horticultural, 30% trees), [b]Surface and drainage weighted average

Table 5. Irrigation requirement for 90% yield (IR_{90}), recommended irrigation (I_{rec}), actual evapotranspiration (ET_a), and drainage (D) all in mm yr^{-1}, EC (dS m^{-1}) and SAR ((mmol L^{-1})$^{1/2}$) of the saturation extract and of the drainage water calculated for drip irrigation

3. Conclusion

Modern irrigation faces a problem of optimization to attain maximum agricultural profitability with minimum damage to natural resources. This demands a precise use of water in the fields, which can be carried out combining i) modelling with ii) monitoring of soil water and salinity, and with iii) irrigation manager or advisor experience. Validated soil salinity models can assist on the development of optimum guidelines for the use of water in salt-threatened areas. The SALTIRSOIL_M model has been developed from the SALTIRSOIL model for the calculation of soil solution major ion composition, pH and electrical conductivity in monthly steps. The time variable has been included in the SALTIRSOIL_M preserving the original capabilities of the SALTIRSOIL model, i.e. maximum reliability-to-data-requirements. In fact no additional data is needed to run SALTIRSOIL_M regarding the original SALTIRSOIL. Just as occurred with the SALTIRSOIL more accurate results can be obtained if detail data on soil layers and monthly water composition is provided to the model. With such simple extension the SALTIRSOIL_M model provides lower leaching requirements. Therefore, similar leaching requirements, and hence, irrigation requirements, to those calculated with more complex transient-state soil salinity models.

The SALTIRSOIL_M model can be used to help develop irrigation guidelines. As such it was used for the important traditional irrigation district of *Vega Baja del Segura* (SE Spain). This is located in the lower basin of the Segura River, which lower reaches are featured by high salinity. This is therefore a salt-threatened area. According to the simulations carried out with some of the most important irrigated crops in the district, irrigation could be indefinitely go on without loss of agricultural profitability and preserving natural water quality and amount providing the following recommendations are observed: i) use of 100 hm^3 yr^{-1} of Segura River water to irrigate the 15000 ha of land in the district, i.e, an average of 670 mm yr^{-1}, ii) use of tolerant rootstocks for citrus growth, iii) replacement of surface by localized irrigation systems, iv) maintenance of the system of canals to dispose of the drainage effluents.

The data from soil water and salinity probes along with the irrigation manager or advisor experience should then be used to precisely adapt such guidelines to the plot and plant scales. Soil salinity models are, therefore, the key factor in the development of decision support systems for the sustainable use of water in irrigated areas.

4. Acknowledgment

The authors would like to acknowledge the *Ministerio de Ciencia e Innovación* from the Government of Spain for funding the projects CGL2009-14592-C02-01 and CGL2009-14592-C02-02 for the development of the SALTIRSOIL_M model. F. Visconti would also like to thank the *Conselleria d' Educació* from the *Generalitat Valenciana* for funding his work through a postdoctoral scholarship in the framework of program VAL i+d 2010.

5. References

Abadía, R.; Ortega, J.F.; Ruíz, A. & García, T. (1999). Analysis of the problems of traditional irrigation in the Vega Baja del Segura I : Current situation and considerations about its modernization. *Riegos y Drenajes XXI* Vol. 15, No.108, pp. 21-31, ISSN 0213-3660 (in Spanish)

Abdelkhaleq, R.A. & Ahmed, I.A. (2007). Rainwater harvesting in ancient civilizations in Jordan, In: *Insights into Water Management: Lessons from Water and Wastewater Technologies in Ancient Civilizations*, A.N. Angelakis, D. Koutsoyiannis (Eds.), 85-93, ISBN: 978-1-84339-610-9, Iraklio, Greece

AQUASTAT (2008). FAO's Information System on Water and Agriculture, 20.07.2011, Available from http://www.fao.org/nr/water/aquastat/main/index.stm

Ayers, R.S. & Westcot, D.W. (1985). Water quality for agriculture. FAO, Rome. 25.04.2011, Available from: http://www.fao.org/DOCREP/003/T0234E/T0234E00.HTM

Bustan, A.; Cohen, S.; De Malach, Y.; Zimmermann, P.; Golan, R.; Sagi, M. & Pasternak, D. (2005). Effects of timing and duration of brackish irrigation water on fruit yield and quality of late summer melons. *Agricultural Water Management* Vol. 74, No.2, pp. 123-134, ISSN: 0378-3774

Corwin, D.L.; Rhoades, J.D. & Simunek, J. (2007). Leaching requirement for soil salinity control: Steady-state versus transient models. *Agricultural Water Management* Vol.90, No.3, pp. 165-180, ISSN: 0378-3774

de Paz, J.M.; Visconti, F. & Rubio, J.L. (2011). Spatial evaluation of soil salinity using the WET sensor in the irrigated area of the Segura river lowland. *Journal of Plant Nutrition and Soil Science* Vol.174, No.1, pp. 103-112, ISSN 1436-8730

Goncalves, M.C.; Simunek, J.; Ramos, T.B.; Martins, J.C.; Neves, M.J. & Pires, F.P. (2006). Multicomponent solute transport in soil lysimeters irrigated with waters of different quality. *Water Resources Research* Vol.42, No.8, Article Number: W08401, ISSN: 0043-1397

Grieve, A.M.; Prior, L.D. & Bevington, K.B. (2007). Long-term effects of saline irrigation water on growth, yield, and fruit quality of Valencia orange trees. *Australian Journal of Agricultural Research* Vol.58, No.4, pp. 342-348, ISSN: 0004-9409

Hoffman, G.J. & Van Genuchten, M. Th. (1983). Soil Properties and Efficient Water Use: Water Management for Salinity Control, In: *Limitations to Efficient Water Use in Crop Production*, H.M. Taylor, W.R. Jordan & T.R. Sinclair, (Eds.), 73-85, American

Society of Agronomy, Crop Society of America, Soil Science Society of America, ISBN 0-89118-074-5, Madison, WI (USA)

Huang, Z.B.; Shan, L.; Wu, P.T. & Zhang, Z.B. (1997). Correlation between rainwater use and agriculture sustainable development of Loess Plateau of China, *Proceedings of the 8th International Conference of Rainwater Catchment Systems, Vols 1 and 2: Rainwater Catchment for Survival*, pp. 1048-1054, IDS Number: BR05T, Tehran, Iran, Apr 25-29, 1997

Ibáñez, V. & Namesny, A. (1992). Quality for irrigation of the waters of mainland Spain: I. The Segura Basin. *Tecnología del Agua* Vol.12, No.104, pp. 26-52, ISSN: 0211-8173 (in Spanish)

Isidoro, D. & Grattan, S.R. (2011). Predicting soil salinity in response to different irrigation practices, soil types and rainfall scenarios. *Irrigation Science* Vol.29, No.3, pp. 197-211, ISSN: 0342-7188

Letey, J.; Hoffman, G.J.; Hopmans, J.W.; Grattan, S.R.; Suarez, D.; Corwin, D.L.; Oster, J.D.; Wu, L. & Amrhein, C. (2011). Evaluation of soil salinity leaching requirement guidelines. *Agricultural Water Management* Vol.98, No.4, pp. 502-506, ISSN: 0378-3774

Maas, E.V., Hoffman, G.J., 1977. Crop salt tolerance—Current assessment. *Journal of Irrigation and Drainage Engineering-ASCE* Vol.103, No.IR2, pp. 115-134, ISSN: 0733-9437

MMA. (1997). Hydrological Plan of the *Segura* Basin. Confederación Hidrográfica del Segura. Ministerio de Medio Ambiente. Available from http://www.chsegura.es/chs/planificacionydma/plandecuenca/documentoscom pletos/(in Spanish)

Narasimhan, T.N. (2009). Groundwater: from mystery to management. *Environmental Research Letters*, Vol. 4, No. 3, Article Number: 035002, ISSN: 1748-9326

Ramos, G. (2000). Incorporation of brackish underground water resources to water management. Southern area of Alicante province. *Industria y Minería*, Vol.339, pp. 30-35, ISSN: 1137-8042 (in Spanish)

Rhoades, J.D. (1974). Drainage for salinity control. In: *Drainage for Agriculture*, van Schilfgaarde, J. (Ed.), 433-461, Agronomy Monograph No. 17. SSSA, Madison (Wisconsin, USA),

Rhoades, J.D. & Merrill, S.D. (1976). Assessing the suitability of water for irrigation: theoretical and empirical approaches. In: *Prognosis of salinity and alkalinity* 25.04.2011 FAO, Rome, pp. 69–109. Available from: http://www.fao.org/docrep/x5870e/x5870e00.htm

Schoups, G.; Hopmans, J.W. & Tanji, K.K. (2006). Evaluation of model complexity and space–time resolution on the prediction of long-term soil salinity dynamics, western San Joaquin Valley, California. *Hydrological Processes* Vol. 20, No.13, pp. 2647-2668, ISSN: 0885-6087

Tanji, K.K. & Kielen, N.C. (2002). Agricultural drainage water management in arid and semi-arid areas. FAO, ISBN 92-5-104839-8, Rome, Italy

Turekian, K.K. (1977). The fate of metals in the oceans. *Geochimica et Cosmochimica Acta*, Vol.41, No.8, pp. 1139-1144, ISSN: 0016-7037

Visconti, F.; de Paz, J.M.; Molina, Mª J. & Sánchez, J. (2011a). Advances in validating SALTIRSOIL at plot scale: First results. *Journal of Environmental Management*, doi:10.1016/j.jenvman.2011.03.020, ISSN 0301-4797

Visconti, F., de Paz, J.M., Rubio, J.L. & Sánchez, J. (2011b). SALTIRSOIL: a simulation model for the mid to long-term prediction of soil salinity in irrigated agriculture. *Soil Use and Management*, Vol.27, No.4, pp. 523-537, ISSN 0266-0032

Visconti, F., de Paz, J.M., Rubio, J.L. & Sánchez, J. (2012). Comparison of four steady-state models of increasing complexity for assessing the leaching requirement in agricultural salt-threatened soils. *Spanish Journal of Agricultural Research* (in press), ISSN 1695-971X

Winpenny, J.T. (2003). Managing water scarcity for water security, FAO, 20.07.2011, Available from http://www.fao.org/ag/agl/aglw/webpub/scarcity.htm

Zribi, W., Faci, J.M., Aragüés, R. (2011). Mulching effects on moisture, temperature, structure and salinity of agricultural soils. *Información Técnica Económica Agraria*, Vol.107, No.2, pp. 148-162, ISSN: 1699-6887 (in Spanish with summary in English)

Guideline for Groundwater Resource Management Using the GIS Tools in Arid to Semi Arid Climate Regions

Salwa Saidi[1], Salem Bouri[1], Brice Anselme[2] and Hamed Ben Dhia[1]
[1]Water, Energy and Environment Laboratory (LR3E), ENIS, Sfax
[2]PRODIG Laboratory, Sorbonne University, Paris
[1]Tunisia
[2]France

1. Introduction

The quality of groundwater is generally under a considerable potential of contamination especially in coastal areas with arid and semi-arid climate like the study area. It is also characterized by intensive agriculture activities, improper disposal of wastewater, and occurrence of olive mills. In addition the intensity of exploitation, often characterized by irrational use, imposes pressures on groundwater reserves. Therefore, there is clearly an urgent need for rapid reconnaissance techniques that allow a protection of groundwater resources of this area.

Groundwater management and protection constitutes an expensive undertaking because of the prohibitive costs and time requirements. To preserve the groundwater resources a simple susceptibility indexing method, based on vulnerability and quality index, was proposed.

The groundwater vulnerability assessment has recently become an increasingly important environment management tool for local governments. It allows for better understanding of the vulnerabilities associated with the pollution of local groundwater sub areas, according to local hydrological, geological or meteorological conditions. The adopted method was specifically developed for groundwater vulnerability DRASTIC method and it is a widely used in many cases of study (Aller et al., 1987; Saidi et al., 2009 and 2011;Rahman, 2008). The DRASTIC model is based on seven parameters, corresponding to the seven layers to be used as input parameters for modeling, including depth to water table (D), recharge (R), aquifer type (A), soil type (S), topography (T), impact of vadose zone (I) and conductivity (C). Vulnerability index is defined as a weighted sum of ratings of these parameters. The quality index calculation procedure, based on the water classification, was introduced to evaluate hydrochemical data.

Therefore the main objective of this study is to propose some water management scenarios by performing the susceptibility index (Pusatli et al., 2009) for drinking and irrigation water. The first objective was to evaluate the susceptibility index. To this end, a combination of both vulnerability and water quality maps has been considered. The second objective was to classify the study area into zones according to each degree of susceptibility and some alternatives to manage the groundwater resources of the Chebba – Mellouleche aquifer were proposed.

A geographic information system (GIS) offers the tools to manage, manipulate process, analyze, map, and spatially organize the data to facilitate the vulnerability analysis. In addition, GIS is a sound approach to evaluate the outcomes of various management alternatives are designed to collect diverse spatial data to represent spatially variable phenomena by applying a series of overlay analysis of data layers that are in spatial register.

2. Study area

The region, object of this study, is the Chebba – Mellouleche aquifer which is situated in the Eastern Tunisia with a total surface of 510 km^2 and a coastline of 51Km (Fig. 1). This region is characterized by a semi-arid climate, with large temperature and rainfall variations. Averages of annual temperature and rainfall are about 19.8°C and 225 mm, respectively (Anon., 2007a). It is known for intensive anthropogenic activities such as industrial and especially agricultural ones which is concentrated in its North east part (Fig. 1).

Both of the aquifer and the vadose zone of the Chebba– Mellouleche region are located in Plio-Quaternary layer system which is constituted mainly by alluvial fan, gravel, sand, silt and clay with high permeability (Saidi et al., 2009). Hence, it results in an easily infiltration of nutrients in the groundwater. The aquifer has an estimated safe yield of 3.24 10^6 m^3/yr, but annual abstraction by pumping from 4643 wells stands at 4.28 10^6 m^3/yr (CRDA, 2005).

The groundwater supply is under threat due to salinisation as salinity measures are generally of 1.5-3 g/l in the majority of the coastal Aquifer, and exceed 6 g/L in the West (Anon., 2007b). For these reasons, a new water management planning is highly required.

3. Methodology

It is noted that an integration of hydrogeological and hydrochemical parameters through the use of the susceptibility index method should be considered as a reliable tool for groundwater quality protection and decision making in this region.

To reach this aim, a variety of GIS analysis and geo - processing framework, which includes: Arc Map, Arc Catalog, Arc Scene and Model Builder of the Arc GIS 9.2 were used (Rahman, 2008).

3.1 Susceptibility index (S$_I$)

The contamination susceptibility index (S$_I$) was calculated by considering the product of the vulnerability index (V$_I$) and the quality index (Q$_I$) using the following equation (Pusatli et al., 2009):

$$S_I = V_I * Q_I \qquad (1)$$

3.1.1 Vulnerability index (V$_I$)

In the present study the DRASTIC method, a standard system for evaluating groundwater pollution potential is used. The DRASTIC model is very used all over the world because the input information required for its application is either readily available or easily

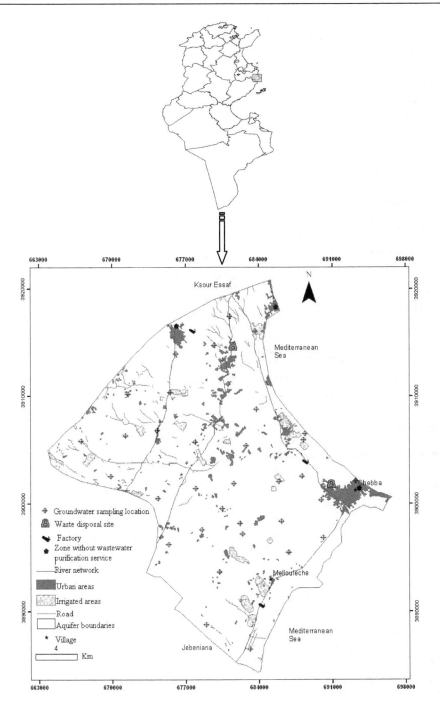

Fig. 1. Location of the study area.

obtained from various government agencies. This model was developed for the purpose of groundwater protection in the United States of America and its methodology is referred as "DRASTIC." This methodology developed as a result of a cooperative agreement between the NWWA and the US Environmental Protection Agency (EPA). It was designed to provide systematic evaluation of GW pollution potential based on seven parameters whose required information were obtained from various Government and semi-Government agencies at a desired scale (Table 1). The acronym DRASTIC stands for the seven hydrogeologic parameters used in the model which are: Depth of water, Net Recharge, Aquifer media, Soil media, Topography, Impact of the vadose zone and hydraulic Conductivity:

Parameter	Data and Sources	Mode of processing
Vulnerability (V_I) or DRASTIC Parameters		
D	Monthly monitoring of shallow wells in 2007 (Anon., 2007b).	Interpolation
R	Precipitation, Evapotranspiration (Anon., 2007a).	Interpolation
A	Geological information (Bedir, 1995), well logs (Anon., 2007).	Interpolation
S	Soil maps (scale 1:50,000) (Anon., 2008).	Digitalization
T	Topographical maps (scale 1:50000) (Anon., 2008).	Digitalization
I	Analysis of water logs and geological maps (Anon., 2007b).	Interpolation
C	Pumping tests (Anon., 2007c).	Interpolation
Quality index (Q_I)	Chemical composition of water wells samples (Anon., 2007c and Trabelsi, 2008)	Interpolation

Table 1. Data sources of susceptibility index (VI and QI) parameters

Depth to groundwater (D): It represents one of the most important factors because it determines the thickness of the material through which infiltrating water must travel before reaching the aquifer-saturated zone. In general, the aquifer potential protection increases with its water depth. The borewell and borehole data was collected from Mahdia Agricultural Agency.

Net Recharge (R): The net recharge is the amount of water from precipitation and artificial sources available to migrate down to the groundwater. Recharge water is, therefore, a significant vehicle for percolating and transporting contaminants within the vadose zone to the saturated zone. To calculate the distribution of the recharge parameter, the water table fluctuations (WTF) method was used. This method estimates groundwater recharge as the product of specific yield and the annual rate of water table rate including the total groundwater draft (Sophocleous, 1991).

Aquifer media (A) and the impact of the vadose zone (I): were represented by the lithology of the saturated and unsaturated zones, which is found in well logs (Saidi et al., 2009).

Topography (T): was represented by the slopes map (1/50 000 scale) covering the study area.

Soil media (S): It considers the uppermost part of the vadose zone and it influences the pollution potential. A soil map, for the study area, was obtained by digitizing the existing soil maps covering the region (Anon., 2008).

Hydraulic conductivity (C): It refers to the ability of the aquifer materials to transmit water, which in turn, controls the rate at which ground water will flow under a given hydraulic gradient. The rate at which the ground water flows also controls the rate at which a contaminant moves away from the point at which it enters the aquifer (Aller et al., 1987).

The hydraulic Conductivity was calculated based on the following equation

$$K = T/b, \qquad\qquad (2)$$

where K is the hydraulic conductivity of the aquifer (m/s), b is the thickness of the aquifer (m) and T is the transmissivity (m²/s), measured from the field pumping tests data.

It is divided into ranges where high values are associated with higher pollution potential. Figure 2 shows the relative importance of the ranges.

Thus, thematic maps representing the D, R, A, I and C parameters were created by interpolation of data used for each one (Table 1). However, the soil type and topography maps are geo-referenced and digitized from different data files (Saidi et al., 2009).

The final vulnerability index is computed as the weighted sum overlay of the seven layers using the following equation:

$$V_I = Dr\ Dw + Rr\ Rw + Ar\ Aw + Sr\ Sw + Tr\ Tw + Ir\ Iw + Cr\ Cw \qquad (3)$$

where D, R, A, S, T, I, and C are the seven parameters and the subscripts r and w are the corresponding rating and weights, respectively.

The DRASTIC vulnerability index was determined from multidisciplinary studies as shown in Table 1. The distributed value of each parameter was the rated in each cell of the grid map of 300 m by 300 m cell dimensions. According to the range of Aller et al. (1987), the contamination vulnerability index was created by overlying the seven thematic layers using intersect function of analysis tools in the Arc Map.

3.1.2 Modification of the weights of the DRASTIC method

The "real" weight is a function of the other six parameters as well as the weight assigned to it by the DRASTIC model (Saidi et al., 2011).

In this analysis real or "effective" weight of each parameter was compared with its assigned or "theoretical" weight. The effective weight of a parameter in a sub-area was calculated by using the following equation:

$$W = ((P_r\ P_w)/V_I)*100 \qquad\qquad (4)$$

where W refers to the "effective" weight of each parameter, P_r and P_w are the rating value and weight for each parameter and V_I is the overall vulnerability index.

3.1.3 Quality index (Q$_I$)

The quality index calculation is based on the quality classes of ions, which were determined using the concentrations of ions in groundwater at a given location. In this application, we used four classification schemes that are described in the following references: WCCR (1991), Anon. (2003), Neubert and Benabdallah (2003) and WHO (2006). In this classification, the irrigation water quality is classified into five groups with respect to each ion concentration as very good (I), good (II), usable (III), usable with caution (IV) and harmful (V). The classification limits used in this study for the considered parameters are listed in Table 2.

1- Irrigation water classification

Parameters	Irrigation water limits				
	Class I (very good)	Class II (good)	Class III (usable)	Class IV (usable with caution)	Class V (harmful)
EC (μS/cm)	0 - 250	250 - 750	750 - 2000	2000 - 3000	> 3000
Cl (mg/l)	0 - 142	142 - 249	249 - 426	426 - 710	> 710
NO_3^- (mg/l)	0 - 10	10 - 30	30 - 50	50 - 100	> 100
SO_4^{2-} (mg/l)	0 - 192	192 - 336	336 - 575	576 - 960	> 960
Na^+ (mg/l)	0 - 69	69 - 200	200 - 252		> 252

2- Drinking water classification

Parameters	Irrigation water limits				
	Class I (very good)	Class II (good)	Class III (usable)	Class IV (usable with caution)	Class V (harmful)
EC (μS/cm)	0 - 180	180 - 400	400 - 2000	2000 - 3000	> 3000
Cl (mg/l)	0 - 25	25 - 200			> 200
NO_3^- (mg/l)	0 - 10	10 - 25	25 - 50		> 50
SO_4^{2-} (mg/l)	0 - 25	25 - 250			> 250
Na^+ (mg/l)	0 - 20	20 - 200			> 200

Table 2. Water classification (WCCR, 1991; Anonymous, 2003; Neubert et Benabdallah, 2003 and WHO, 2006)

The quality index at a given location can be calculated using the following formulation:

$$Q_I = P(C_i)^2 \qquad (5)$$

where summation is overall considered quality parameters (ions). C is the determined class of parameter, i (ion), as an integer number (from 1 to 5) at a given location. The second power of C was used to enhance the effect of poor quality classes in the index (Saidi et al., 2009). In order to determine the chemical composition of the Chebba– Mellouleche groundwater during the irrigation period, 33 samples were collected from wells and analyzed in July 2007 (Saidi, 2011) (Fig. 1). Groundwater samples were taken from 27 wells of the Chebba – Mellouleche Aquifer.

3.2 Water management propositions

The builder model, describing the methodology applied to assess the water susceptibility index, was created using the Arc Tool Box in Arc Map interface of Arc GIS 9.2 (Saidi et al.,

2009). Next, it is possible to propose a management plan by overlying the susceptibility index maps for irrigation and drinking water.

4. Results and discussions

4.1 Modification of the DRASTIC weights

The "real" or effective weights of the DRASTIC parameters exhibited some deviation from the "theoretical" weight (Table 3). The depth to groundwater table and the Aquifer media seem to be the most effective parameters in the vulnerability assessment; The depth of groundwater, D, with an average weight of 20.3% against a theoretical weight of 21.7% assigned by DRASTIC and the Aquifer media parameter, A (25.3%) against a theoretical weight of 13%. The net Recharge, R, the hydraulic conductivity, C, and especially the impact of the vadose zone, I, reveal lower "effective" weights when comparing with the "theoretical" weights.

Parameter	Theoretical weight	Theoretical weight (%)	Effective weight (%)				Real weight after rescaling
			Mean	Minimum	Maximum	SD	
D	5	21.7	20,3	5	36	5,92	4.66
R	4	17.4	10,5	3	25	6,32	2,4
A	3	13	25,3	2	41	3.92	5,81
S	2	8.7	8	0	20	3.93	1,83
T	1	4.4	8.5	1	14	2,14	1.95
I	5	21.8	17	5	24	2,87	3,91
C	3	13	10,5	4	17	2,19	2,41

SD: standard deviation.

Table 3. Statistics of single parameter sensitivity analysis and a comparison between "theoretical" weight and "effective" weight.

4.2 Aquifer vulnerability

The vulnerability map shows three classes as indicated in Fig. 3. The highest class of vulnerability (140–159) covers 25% of the total surface. In fact, zones with high vulnerability correspond to the shallow groundwater table (<9 m), a flat topography (<5%), a high recharge and a permeable lithologies of the vadose zone and The Aquifer (made up of sand and gravel lithology). It results in a low capacity to attenuate the contaminants.

The areas with moderate to low vulnerability cover the rest of the study area, characterized by a deep groundwater table (> 25 m), low recharge (>150 mm) and lithology with low permeability (Table 4).

Using real weights, the high vulnerability class covers the whole of the southern part of the study area. It corresponds to the location of the irrigated areas, using intensive fertilizers. So, the utilization of the calculated or real weights can better reflect the pollution state of the study area than using theoretical weights, in groundwater vulnerability assessment. Therefore, the use of real weights in the DRASTIC index shows more similarity when comparing vulnerability degree and nitrate distribution (Figs. 3).

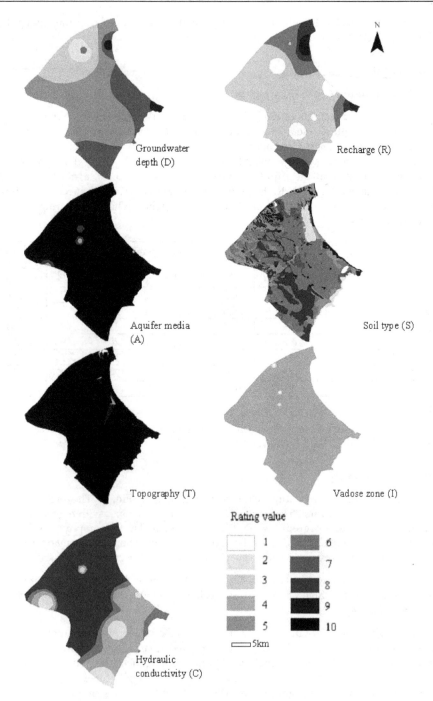

Fig. 2. Seven DRASTIC maps to compute the vulnerability index.

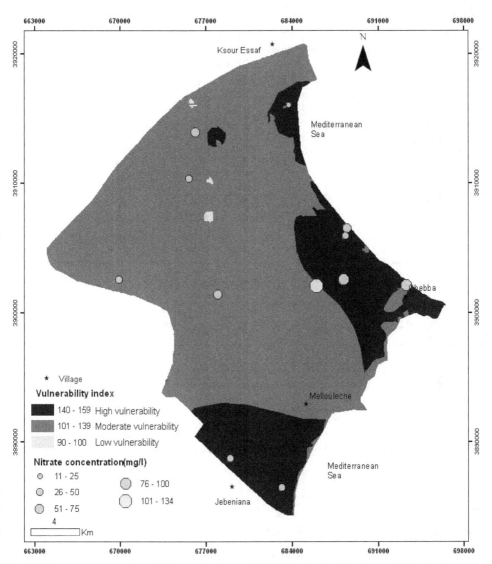

Fig. 3. Groundwater vulnerability and nitrate distribution in the Chebba – Mellouleche Aquifer using DRASTIC method (Saidi, 2011).

4.3 Water quality

Both the drinking and the irrigation water quality present a low quality, especially in the south of the Aquifer (Fig. 5). The main causes are the high permeability of its lithology as well as its localization in the vicinity of an irrigated area with intensive use of fertilizers. There is no similarity between vulnerability classes and water susceptibility classes. Thus, this proves the impact of the irrigation water quality on the aquifer groundwater quality.

Depth of water (m)		Net recharge (m)		Topography (slope) (%)		Hydraulic Conductivity (m/s)		Aquifer media		Impact of the vadose zone		Soil media	
Interval	R	Interval	R	Interval	R	Interval	R	Lithology classes	R	Lithology classes	R	Soil classes	
2-4.5	9	0.01-0.05	1	0-3%	10	$8.3*10^{-5}$ - $4*10^{-5}$	2	Sand and clay	1	confined Aquifer	1	Mineral soil	9
4.5-9	7	0.05-0.10	3	3-5%	9	$4*10^{-5}$ - $2.5*10^{-4}$	4	Massive clay and sand	2	Sandy clay and calcareous	2	Isohumic chestnut soil	8
9-15	5	0.10-0.18	6	5-10%	5	$2.5*10^{-4}$ - $4*10^{-4}$	6	Sand, gravel and clay	4	sand and silt	4	Rendzina	7
15-23	3	0.18-0.25	8	10-15%	3			Sandy gravel	8	Gravel and sand	10	Calcareous brown soil	6
23-32	2	>0.25	9					Gravel and Sand	10			Soil with little evolution	5
												Polygenetic soil	4
												Gypsum soil	3
												Halomorphic soil	2
												Urbain zones	1

R; Rank

Table 4. Ranks of the seven DRASTIC parameters (Aller et al., 1987).

For instance, the extreme North East part of the Aquifer has a high and a moderate vulnerability but a high water quality (low index). As a consequence, this area reveals a low water susceptibility index (Fig. 6). Nevertheless, the centre of the Aquifer which presented a low water quality and moderate vulnerability corresponds to a moderate water susceptibility index. This is due to the high permeability in this area which can cause a rapid infiltration of contaminant from the surface to the groundwater. But, in the South east a high vulnerability and a moderate to low water quality and the results are a moderate to low susceptibility index. The main reasons are probably the lithology of unsaturated zone and the comportment of the contaminants, in this area, which need further investigations (Saidi et al., 2009). The comparison between irrigation and drinking water maps show a few differences; the drinking water indexes are stricter than the irrigation ones (Fig. 6).

According to the drinking water susceptibility index map, people can exploit only the Northern part of the Aquifer for drinking uses and for irrigation of sensible plants. This is due to the high capacity of the unsaturated zone to attenuate the contaminant infiltration (made up of silt, clay and sandy clay) and the deep groundwater table in this area (>25 m) (Fig.2).

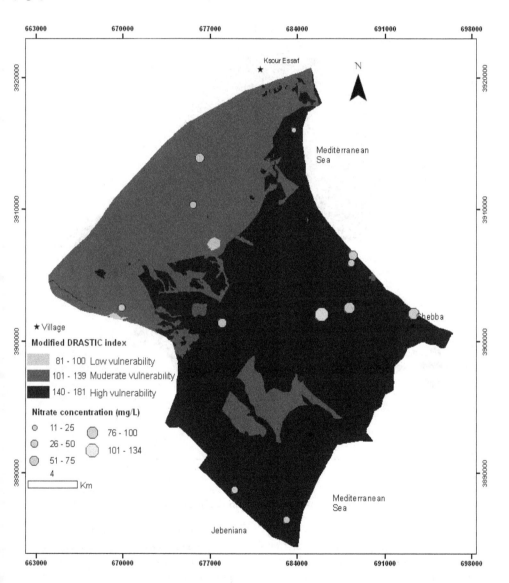

Fig. 4. Groundwater vulnerability and nitrate distribution in of the Chebba – Mellouleche Aquifer using modified DRASTIC method.

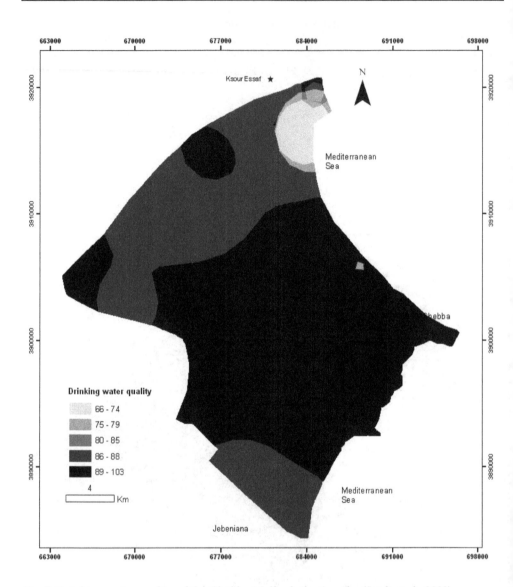

Fig. 5. Drinking water quality of the Chebba-Mellouleche Aquifer (Saidi et al., 2009).

However the Southern part of the study area presents a low water quality because it coincides with a variety of sites and activities which are hazardous to groundwater such as waste disposal sites (which have no technical or geological barrier), industrial estates (which have no proper sewage treatment facilities), agriculture (which applies fertilizers and pesticides abundantly) and fish farming in the vicinity of the coast (where antibiotic and pesticides are used in abundance and imports saltwater increases the salinity in the surrounding area) (Fig.6).

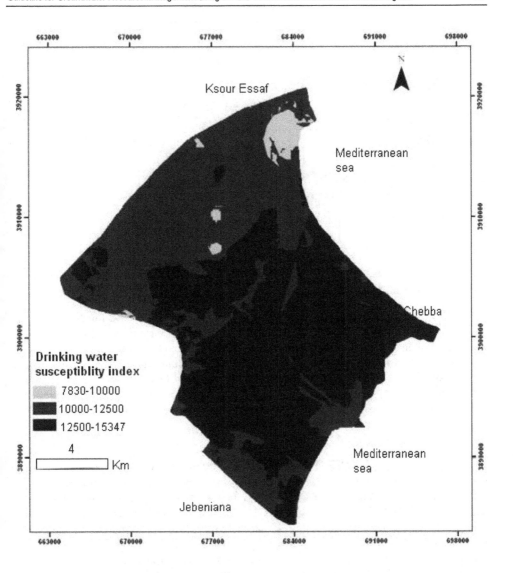

Fig. 6. Drinking water susceptibility index of the Chebba – Mellouleche Aquifer.

The Builder Model, created for the susceptibility indexing assessment, displays and provides a description of the procedures and the geo-processing operations which are used to create the susceptibility index maps (Fig. 7).

So, it can help to retain the main tools for the susceptibility assessment used in this study and facilitate the proposition of a water management schema (Saidi et al., 2009). In fact, a management map was created by overlaying the susceptibility index maps for irrigation and drinking water (Fig. 8).

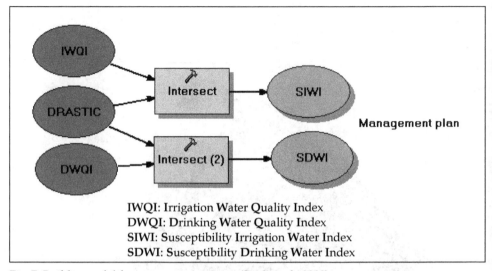

Fig. 7. Builder model for water management (Saidi et al., 2009).

This map shows that: (i) in the north of the aquifer the water can be used for drinking water and for irrigation of the sensible crops (ii) in the Extreme north eastern corner, the water has a high water quality but it represents a high risk since it is near to the coast. So, we should allow additional wells to ovoid seawater intrusion (iii) in the southern part of the study area, we should not allow additional high risk activities in order to obtain economic advantage and reduce environmental pollution hazard. Furthermore, water should be decontaminated before applying to reduce diseases to sensitive plants and should not be utilized for drinking uses.

5. Conclusions

The use of both intrinsic vulnerability data and quality one in a GIS environment proved to be a powerful tool for the groundwater management in arid and semi arid regions like Chebba–Mellouleche. The seven DRASTIC parameters: depth of groundwater, net recharge, aquifer media, soil media, topography, impact of the vadose zone and hydraulic conductivity, were used to calculate the vulnerability of the study area. The results show that groundwater in Chebba – Mellouleche is characterized by four classes as follow: Moderate vulnerability ranked groundwater areas dominated the study area (>52%), which occupy middle of the study area, while (>38%) of the Chebba – Mellouleche aquifer is under high groundwater vulnerability.

The water susceptibility indexes show a low water quality, covering the majority of the study area. Indeed, there is a high similarity between the more hazardous pollution zones and the areas with low water quality. So, these scenarios proposed by this study could be used as a general guide for groundwater managers and planners.

The GIS technique has provided an efficient environment for analyses and high capabilities in handling a large quantity of spatial data. The susceptibility index parameters were constructed; classified and mapped employing various map and attribute GIS functions.

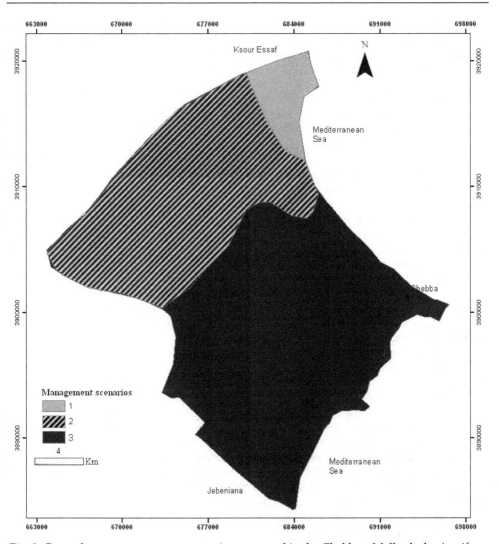

Fig. 8. Groundwater management scenarios proposed in the Chebba – Mellouleche Aquifer.

6.References

Aller, L., Bennet, T., Lehr, J.H., Petty, R.J., Hacket, G., 1987. DRASTIC: a standardized system for evaluating ground water pollution potential using hydrogeologic settings. US Environmental Protection Agency Report (EPA/600/2-87/035), Robert S. Kerr Environmental Research Laboratory, 455 pp.

Anon., 2007a. Tableaux climatologiques mensuels, stations de Mahdia (unpubl.). Annuaire de l'Institut de la Météorologie Nationale. Tunisia.

Anon., 2007b. Annuaires de surveillance de piézométrie, de nitrate et de salinité (unpubl.). Arrondissement des Ressources en Eaux de Mahdia. CRDA, Mahdia, Tunisia.

Anon., 2007c. Comptes rendus des forages et piézomètres de surveillance (unpubl.). Arrondissement des Ressources en Eaux deMahdia. CRDA,Mahdia,Tunisia

Anon., 2008. Carte numérique agricole de Mahdia 1:50,000.Official numerical file. at: http://www.carteagricole.agrinet.tn/GIS/33_MAHDIA/33_mahdia_fcarto/viewe r.htm.

Bedir, M., 1995. Mécanismes géodynamiques des bassins associés aux couloirs de coulissement de la marge atlasique de la Tunisie: seismo- stratigraphie, seismo-tectonique et implications pétrolières. Thèse de doctorat ES- SCIENCES. Université de Tunis II, Tunisie, 414 pp.

Civita, M., Chiappone, A., Falco, M., Jarre, P., 1990. Preparazione della carta di vulnerabilita per la rilocalizzazione di un impianto pozzi dell'acquedutto di Torino. Proc. Ist. Conv. Naz. Protezione e Gestione del Vulnerability validation le acque sotterane: metodologie. Tecnologie e Obiettivi. Morano sol Panero. 2, 461–462.

CRDA Mahdia (Commissariat Régional du Développement Agricole de Mahdia), 2005. Annuaires d'exploitation des nappes phréatiques du gouvernorat de Mahdia.

Foster, S., Hirata, R., 1991. Groundwater pollution risk assessment. A methodology using available data; World Health Organisation (WHO)/Pan American Health Organisation (PAHO)/Pan American Center for Sanitary Engineering and Environmental Sciences (CEPIS), Technical Report, Lima, Perou, 2nd edition, 73 pp.

Neubert and Benabdallah., 2003. Système d'évaluation de la qualité des eaux souterraines SEQ - Eaux Souterraines. Rapport de présentation, version 0.1- Aout 2003, 75 pp.

Pusatli, T.O., Camur, Z.M., Yazicigil, H., 2009. Susceptibility indexing method for irrigation water management planning: applications to K. Menderes river basin, Turkey. J. Environ. Manage. 90, 341–347.

Rahman, A., 2008. A GIS based DRASTIC model for assessing groundwater vulnerability in shallow Aquifer in Aligarh, India. Appl. Geogr. 28, 32–53.

Saidi, S., Bouri, S., Ben Dhia, H., Anselme, B., 2009. A GIS-based susceptibility indexing method for irrigation and drinking water management planning: Application to Chebba–Mellouleche aquifer, Tunisia. Agricultural Water Management 96, 1683–1690.

Saidi, S., 2011.Contribution des approches paramétriques, cartographiques et statistiques à l'étude de la vulnérabilité du système aquifère phréatique de Mahdia. Thèse de doctorat. Université de Sfax, Tunisie, 254 pp.

Saidi, S., Bouri, S., Ben Dhia, H., 2011. Sensitivity analysis in groundwater vulnerability assessment based on GIS in the Mahdia-Ksour Essaf aquifer, Tunisia: a validation study. Hydrol. Sci. J. 56(2), 288–304.

Sophocleous, M.A., 1991. Combining the soil water balance and water level fluctuation methods to estimate natural groundwater recharge: practical aspects. Hydrol. J. 124, 229–241.

Trabelsi, R., 2008. Contribution a l'étude de la salinisation des nappes phréatiques côtières, cas du système de Sfax -Mahdia. Thèse de doctorat Université de Sfax, Tunisie 198 pp.

Van Stempvoort D., Ewert L., Wassenaar L.,1992. Aquifer vulnerability index: a GIS-compatible method for groundwater vulnerability mapping. Canadian Water Resources Journal, 18: pp 25-37.

WCCR (Water Contamination Control Regulations), 1991. Official Paper. Ankara, No. 19919.

World Health Organisation (WHO), 2006.. Guidelines for Drinking-water Quality. Recommendations, 3rd edition, vol. 1. WHO, ISBN 92 4 154696 4, 595 pp.

3

Precision Irrigation: Sensor Network Based Irrigation

N. G. Shah[1] and Ipsita Das[2]
[1]CTARA, IIT Bombay
[2]Department Of Electrical Engg, IIT Bombay
India

1. Introduction

Availability of water for agriculture is a global challenge for the upcoming years. This chapter aims at describing components of precision irrigation system and its potential in future farming practices. A case-study of deploying Wireless Sensor Network (WSN) monitoring soil moisture, estimating Evapotranspiration (ET) and driving drip irrigation for a large grape farm in India on pilot basis is described.

Agriculture plays a vital role in the economy of every nation be it developing or developed. Agriculture is the basis of livelihood for the population through the production of food and important raw materials. Moreover, agriculture continues to play an important role in providing large scale employment to people. Agricultural growth is considered necessary for development and for a country's transformation from a traditional to a modern economy. Therefore attention must be accorded to science and technology being used in the field for higher yield and growth in agriculture.

Agriculture system is a complex interaction of seed, soil, water, fertilizer and pesticides etc. Optimization of the resources is essential for sustainability of this complex system. Unscientific exploitation of agricultural resources to bridge the gap in supply/demand owing to the population growth is leading to resource degradation and subsequent decline in crop yields (Mondal and Tiwari, 2007). In addition, uncertainty of climatic conditions is also playing an important role in this complex system. This calls for optimal utilization of the resources for managing the controlled agricultural system (Whelan et al., 1997). Also agricultural systems are inherently characterized by spatial and temporal variability making yield maximization with minimal inputs a complex task. Thus the farming technologies followed in all parts of world need to be constantly updated to meet these challenges. Development of a range of new technologies in different parts of the world has brought agriculture to a whole new level of sophistication. In fact, modern agriculture has already undergone a sea-change from the ancient times. The concept of precision agriculture has been around for some time now. A new approach of collecting real time data from the environment could represent an important step towards high quality and sustainable agriculture. Precision agriculture is an agricultural system that can contribute to the sustainable agriculture concepts.

1.1 Precision Agriculture (PA)

The term "precision agriculture" is defined as the application of various technologies and principles to manage spatial and temporal variability associated with all aspects of agricultural production (Pierce and Nowak, 1999). Conventional agronomic practices always follow a standard management option for a large area irrespective of the variability occurring within and among the field. For decades now, the farmers have been applying fertilizers based on recommendations emanating from research and field trials under specific agro-climatic conditions (Ladha et al., 2000). Applications of agricultural inputs at uniform rates across the field without taking in-field variations of soil and crop properties into account, does not give desirable crop yield. Consideration of in-field variations in soil fertility and crop conditions and matching the agricultural inputs like seed, fertilizer, irrigation, insecticide, pesticide, etc. in order to optimize the input or maximizing the crop yield from a given quantum of input, is referred to as Precision Agriculture (PA). It is an information-based and technology-driven agricultural system, designed to improve the agricultural processes by precisely monitoring each step to ensure maximum agricultural production with minimized environmental impact. It involves the adjustment of sowing parameters, the modulation of fertilizers doses, site-specific application of water, pesticide and herbicides, etc (Adams et. al., 2000). Irrigating farms backed-up by estimated water-requirements is one of the essential components of precision agriculture to reduce water wastage. Given the limited water resources, optimizing irrigation efficiency is very essential.

1.2 Precision irrigation

Water plays a crucial role in photosynthesis and plant nutrition. The problem of agricultural water management is today widely recognized as a major challenge that is often linked with development issues. Agriculture consumes 70% of the fresh water i.e. 1,500 billion m^3 out of the 2,500 billion m^3 of water is being used each year (Goodwin and O'Connell 2008). It is also estimated that 40% of the fresh-water used for agriculture in developing countries is lost, either by evaporation, spills, or absorption by the deeper layers of the soil, beyond the reach of plants' roots (Panchard et. al., 2007). Post green-revolution era agriculture in India is facing a technological fatigue for two reasons viz a) high rates of ground-water depletion and b) soil salinity due to excessive irrigation in some pockets. Efficient water management is a major concern in many crop systems. More and more planners as well as farmer associations are becoming conscious about warder-audit and water utilization efficiency as the water resources is getting more and more scarce. Efforts of using micro-irrigation methods such as sprinkler and drip irrigation have been made in last three decades in many parts of the world. It has been reported that in year 2005, 1.15 million ha was under micro-irrigation (drip and sprinkler) in India (Modak, 2009). There is no ideal irrigation method available which may be suitable for all weather conditions, soil structure and variety of crops cultures. In the semi-arid areas of developing countries, marginal farmers and small farmers (with a land holding between 2 and 4 hectares) who cannot afford to pay for powered irrigation, heavily depend on the rainfall for their crops. It is observed that farmers have to bear huge financial loss because of wrong prediction of weather and incorrect irrigation method. In light of a real need to improve the efficiency of irrigation systems and prevent the misuse of water, the focus is to develop an intelligent irrigation scheduling system which will enable irrigation farmers to optimize the use of water and only irrigate where and when need for as long as needed.

Precision irrigation is worldwide a new concept in irrigation. Precision irrigation involves the accurate and precise application of water to meet the specific requirements of individual plants or management units and minimize adverse environmental impact. Commonly accepted definition of Precision irrigation is sustainable management of water resources which involves application of water to the crop at the right time, right amount, right place and right manner thereby helping to manage the field variability of water in turn increasing the crop productivity and water use efficiency along with reduction in energy cost on irrigation. It utilizes a systems approach to achieve 'differential irrigation' treatment of field variation (spatial and temporal) as opposed to the 'uniform irrigation' treatment that underlies traditional management systems.

1.2.1 Benefits of precision irrigation

Precision irrigation has the potential to increase both the water use and economic efficiencies. It has been reported that precision irrigation (Drip and Sprinkler) can improve application efficiency of water up to the tune of 80-90% as against 40-45% in surface irrigation method (Dukes, 2004). Results from case studies of variable rate irrigation showed water savings in individual years ranging from zero to 50%. The potential economic benefit of precision irrigation lies in reducing the cost of inputs or increasing yield for the same inputs.

1.2.1.1 Water savings

The primary goal of precision irrigation is to apply an optimum amount of irrigation throughout fields. It is reported by many researchers as the most likely means of achieving significant water savings (Evans and Sadler, 2008). The site specific or variable rate irrigation is considered as a necessary or essential component of precision irrigation. Most researchers expect a reduction in water use on at least parts of fields, if not a reduction in the value aggregated over entire fields (Sadler et al. 2005). It has been reported that variable rate irrigation could save 10 to 15% of water used in conventional irrigation practice (Yule et al. 2008). Hedley and Yule (2009) suggested water savings of around 25% are possible through improvements in application efficiency obtained by spatially varied irrigation applications.

1.2.1.2 Yield and profit

The experimental studies were carried out by King et al. (2006) for measuring the yield of potatoes under spatially varied irrigation applications. It was reported that yields were better in two consecutive years over uniform irrigation management. Booker et al. (2006) analyzed yields and water use efficiency for spatially varied irrigation over four years for cotton. They concluded that cotton seems to be unpredictable to manage with spatially varied irrigation. This result is supported by the work of Bronson et al. (2006).

1.3 Components of precision irrigation system

a.   **Data acquisition**

A Precision Irrigation system requires ability to identify and quantify the variability i.e. spatial and temporal variability that exist in soil and crop conditions within a field and between fields. Existing technology is available to measure the various components of the soil-crop-atmosphere continuum many in real-time so as to provide precise and/or real-time control of irrigation applications.

b.   Interpretation

Data has to be collected, interpreted and analyzed at an appropriate scale and frequency. The inadequate development of decision support systems has been identified as a major bottle neck for the interpretation of real time data and adoption of precision agriculture (McBratney *et al.*, 2005).

c.   Control

The ability to optimize the inputs and adjust irrigation management at appropriate temporal and spatial scales is an essential component of a precision irrigation system. Applying differential depths of water over a field will be dependent on the irrigation system. Automatic controllers with real time data should provide the most reliable and accurate means of controlling irrigation applications.

1.4 Technology associated with precision irrigation

The advent of precision irrigation methods has played a major role in reducing the quantity of water required in agricultural and horticultural crops, but there is a need for new methods of automated and accurate irrigation scheduling and control. The early adopters found precision agriculture to be unprofitable and the instances in which it was implemented were few and far between. Further, the high initial investment in the form of electronic equipment for sensing and communication meant that only large farms could afford it. The technologies used are *Remote Sensing* (RS), *Global Positioning System* (GPS) and *Geographical Information System* (GIS) and *Wireless Sensor Network* (WSN).

The technology of GIS and GPS apart from being non-real-time, involved the use of expensive technologies like satellite sensing and also labor intensive. Over the last several years, the advancement in sensing and communication technologies has significantly brought down the cost of deployment and running of a feasible precision agriculture framework. However, a stand-alone sensor, due to its limited range, can only monitor a small portion of its environment but the use of several sensors working in a network seems particularly appropriate for precision agriculture. The technological development in Wireless Sensor Networks made it possible to monitor and control various parameters in agriculture. Also recent advances in sensor and wireless radio frequency (RF) technologies and their convergence with the Internet offer vast opportunities for application of sensor systems for agriculture. Emerging wireless technologies with low power needs and low data rate capabilities have been developed which perfectly suit precision agriculture (Wang et al., 2006). The sensing and communication can now be done on a real-time basis leading to better response times. The wireless sensors are cheap enough for wide spread deployment and offer robust communication through redundant propagation paths (Akyildiz & Xudong, 2005). The wireless sensor networks (WSNs) have become the most suitable technology to monitor the agricultural environment.

1.4.1 Wireless Sensor Network (WSN)

Wireless sensor networks (WSN) is a network of small sensing devices known as sensor nodes or motes, arranged in a distributed manner, which collaborate with each other to gather, process and communicate over wireless channel about some physical phenomena. The sensor motes are typically low-cost, low-power, small devices equipped with limited sensing, data processing and wireless communication capabilities with power supply, which perfectly suites the PA/PI (Wang, 1998; Stafford, 2000). A wireless sensor is a self-powered computing unit usually containing a processing unit, a trans-receiver and both analog and digital interfaces, to which a variety of sensing units such as temperature, humidity etc. can be adapted (Fig 1.1). The sensor nodes communicate with each other in order to exchange and process the information collected by their sensing units. If nodes communicate only directly with each other or with a base station, the network is single-hop. In some cases, nodes can use other wireless sensors as relays, in which case the network is said to be multi-hop. In a data-collection model, sensors communicate with one or several base stations connected to a database and an application server that stores the data and performs extra data-processing. The result is available typically via a web-based interface.

1.4.1.1 Wired vs. Wireless Network

Wireless sensor network have a big potential for representing the inherent soil variability present in fields with more accuracy than the current systems available. WSN can operate in a wide range of environments and provide advantages in cost, size, power, flexibility and distributed intelligence, compared to wired ones. The wireless sensors are cheap enough for wide spread deployment in the form of a mesh network and also it offers robust communication through redundant propagation paths (Roy et al., 2008).

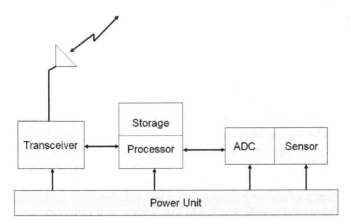

Fig. 1.1. Depiction of Sensor Node

The advantage for wireless sensor network over wired one is the feasibility of installation in places where cabling is impossible. Another obvious advantage of wireless transmission over wired one is the significant reduction in cost and simplification in wiring and harness (Akyildiz et al., 2002). It has been reported that adopting wireless technology would eliminate 20-80% of the typical wiring cost in industrial installations (Wang et al., 2006). However, wired networks are very reliable and stable communication systems for instruments and control. Since installation of WSN is easier than wired network, sensors can be more densely deployed to provide local detailed data necessary for precision agriculture.

Another advantage is their mobility i.e. sensors can be placed on rotating equipment, such as a shaft to measure critical agricultural and environmental parameters. Whenever physical conditions change rapidly over space and time, WSNs allow for real-time processing at a minimal cost. Their capacity to organize spontaneously in a network makes them easy to deploy, expand and maintain, as well as resilient to the failure of individual measurement points. Over the last few years, the advancement in sensing and communication technologies has significantly brought down the cost of deployment and running of a feasible precision agriculture using WSN (Wang, 1998).

Wireless sensor network (WSN), a potential technology found to be suitable for collecting real time data for different parameters pertaining to weather, crop and soil helps in developing solutions for majority of the agricultural processes related to irrigation and other agricultural processes. The development of wireless sensor applications in agriculture makes it possible to increase efficiency, productivity and profitability of farming operations.

2. Irrigation scheduling through evapo-transpiration

Irrigation scheduling defined by Jensen (1981) is as "a planning and decision-making activity" that the farm manager is involved in before and during most of the growing season for each crop that is grown." In other words it is a process through which water lost by the plant through the evapo-transpiration (ET) method is an excellent way to determine how much water to apply based on estimates of the amount of water lost from the crop. Water use efficiency can be achieved with the precisely scheduled irrigation plan. Such a plan on daily basis provides a means of irrigating with an exact amount of water at the targeted dry area to fulfill the needs of evapo-transpiration (ET).

2.1 Evapo-transpiration / crop water requirement (ET)

The combination of two separate processes whereby water is lost on the one hand from the soil surface by evaporation and on the other hand from the crop by transpiration is referred to as evapo-transpiration (ET). Evapo-transpiration is also known as water requirement of the crops. The water requirement can be supplied by stored soil water, precipitation, and irrigation. Irrigation is required when ET (crop water demand) exceeds the supply of water from soil water and precipitation. As ET varies with plant development stage and weather conditions, both the amount and timing of irrigation are important. The rate of ET is a function of four critical factors i.e. weather parameters, soil moisture, plant type and stage of development (Allen *et al.*, 1998). Different crops have different water-use requirements under the same weather conditions. The evapo-transpiration rate from a reference surface is called the reference crop ET and denoted as ET_0. The reference sureface is hypothetical grass reference crop with an assumed crop height of 0.12 m, a fixed surface resistance of 70 sec m^{-1} and an albedo (reflectance of the crop-soil surface i.e. fraction of ground covered by vegetation) of 0.23, closely resembling the evapo-transpiration from an extensive surface of green grass of uniform height, actively growing, well-watered, and completely shading the ground" (Allen et al., 1989). The grass is specifically defined as the reference crop. The crop coefficients appropriate to the specific crops are used along with the values of reference ET for computing the actual ET at different growth stages of the crop. The modified Penman and Moneith model (shown in equation 2.1) was used to calculate the reference evapotranspiration. The calculation procedures of ET_0 by means of the FAO Penman-Monteith equation (Eq. 2.1) are presented by Allen et al (1998).

$$ET_0 = \frac{0.408 \; \Delta(R_n - G) + \gamma \dfrac{900}{T + 273} U_2(e_s - e_a)}{\Delta + \gamma(1 + 0.34 U_2)}$$ (2.1)

Where:

ET_0	Reference evapo-transpiration [mm day^{-1}],
R_n	Net radiation at the crop surface [MJ m^{-2} day^{-1}],
G	soil heat flux density [MJ m^{-2} day^{-1}],
T	Air temperature at 2 m height [°C],
U_2	Wind speed at 2 m height [m s^{-1}],
e_s	Saturation vapour pressure [kPa],
e_a	Actual vapour pressure [kPa],
$e_s - e_a$	Saturation vapour pressure deficit [kPa],
Δ	Slope of vapour pressure curve [kPa °C^{-1}],
γ	Psychrometric constant [kPa °C^{-1}].

2.1.1 Actual crop Evapo-transpiration (ET$_c$)

The crop evapo-transpiration differs distinctly from the reference evapo-transpiration (ET$_0$) as the ground cover, canopy properties and aerodynamic resistance of the crop are different from grass. The Kc component of equation integrates the characteristics of the crop (e.g., crop height, fraction of net radiation absorbed at the land surface, canopy resistance, and evaporation from bare soil surface) into the ETc estimation equation, to account for the difference in transpiration between the actual crop and the reference grass. The effects of characteristics that distinguish field crops from grass are integrated into the crop coefficient (K$_c$). In the crop coefficient approach, crop evapo-transpiration is calculated by multiplying ET$_0$ by K$_c$.

$$ET_c = K_c \times ET_0.$$ (2.2)

Where:

ET_c	Crop evapo-transpiration [mm day^{-1}],
K_c	Crop coefficient [dimensionless],
ET_0	Reference crop evapo-transpiration [mm day^{-1}].

2.2 A case study on using Wireless Sensor Network (WSN) in estimating crop water requirement at Sula vineyard, Nashik, India

Grapes cultivation in India is limited due to high recurring cost of cultivation. There is significant variability in the quality of grapes over the years and also within the field. Assessing the yield and quality (both temporal and spatial) is a big challenge for wineries (Das et. al., 2010). Vine soil-water status constitutes one of the main driving factors which affect plant vegetative growth, yield and wine test and quality. Irrigation requirements are currently estimated from winter/summer season as well as berry forming stages. Providing the methods and tools for continuous measurement of soil and crop parameters to characterize the variability of soil water status will be of great help to the grape growers. A wireless sensor network can facilitate creation of a real-time networked database. The real time information from the fields such as soil water content, temperature, and plant characteristics provided a good base for making decisions on irrigation i.e. (when and how much water to apply). The objective of our study was to relate irrigation requirement through evapo-transpiration. The section below describes the agricultural experiments conducted in the grape field which concentrated on monitoring different parameters relating to crop, soil and climate by deploying the wireless sensors network so as to establish a correlation between sensors output and agricultural requirement in terms of water management.

2.2.1 Experimental setup at Green House, IIT Bombay and vineyard at Nashik

Initial deployment of sensors with a wireless sensor network (WSN) in a greenhouse at IIT Bombay (6 X 9 m) provided a pilot scale crop monitoring environment. It was used for testing the ruggedness of WSN for crops grown under controlled conditions in a greenhouse, using sensors embedded in soil and surrounding which was later extended to a larger scale in an intensely cultivated commercial grape farm i.e. Sula vineyard at Nashik (India). Initially the WSN was tested in a greenhouse of 6 X 9 m in the laboratory at Indian Institute of Technology-Bombay (India). Okra plants were planted in nine plots (1.5 X 3 m), with four plants in a row, maintaining a distance between rows and plant of 50 and 30 cms respectively. WSN system deployed consisted of the battery-powered nodes equipped with sensors for continuously monitoring agricultural parameters consisting of air temperature, air relative humidity, soil temperature and soil water content. These parameters were periodically monitored and transmitted in a multi-hop to a centralized processing unit (see section 2.2.2). The measured and recorded values of parameters in real time over a period of 3 months permitted the calculation of evapotranspiration (ET) (Shah et al., 2009). Figure 2.1 shows the schematics of agricultural environment sensors deployed in the field while Fig 2.2 shows the sensors deployed in greenhouse, IIT Bombay.

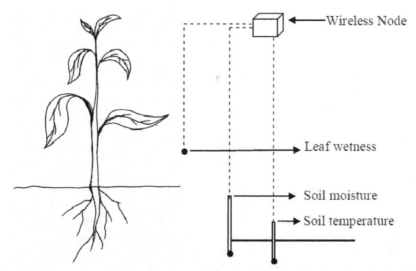

Fig. 2.1. Schematics of Agricultural Environmental Sensors deployed in the Field

Fig. 2.2. Deployment of Wireless sensor network (WSN) System

The WSN system tested at IIT- Lab facility was extended to Sula Vineyard, Nashik (India), for grape crop monitoring as shown in Figure 2.3. The sensors were deployed at a grid of 30 m by 30 m. Each node was able to transmit/receive packets to other nodes inside a well-defined transmission range of 30m. WSN system was focused on establishing feasibility of capturing and analyzing data and facilitated global data accessibility from a small number of wireless sensor pods.

Fig. 2.3. Wireless Sensor Network Deployment at Vineyard, Nashik, MH, India

2.2.2 Details of developed Wireless Sensor Network at IIT Bombay

The system designed, developed and deployed at IIT Bombay for its utility for in-field monitoring of grape crop performance, is being popularly known as AgriSens. (Das et. al.,

2010). It used a combination of wired and wireless sensors to collect sensory data such as soil pH, soil moisture, soil temperature, etc. Data collected by the sensors were wirelessly transferred in multi-hop manner to a base station node (about 700 m away from the mote) connected with embedded gateway for data logging and correlation. An embedded gateway base station performed elementary data aggregation and filtering algorithms and transmitted the sensory data to Agri-information server via GPRS, a long distance, high data-rate connectivity as illustrated in Fig 2.4. Here the data was processed and stored in a structured database to provide useful information to the farmers to take action such as, e.g., starting or stopping of the irrigation system. The server was situated at Signal Processing Artificial Neural Network lab, Department of Electrical Engineering IIT Bombay, India which is about 200 km away from the fields. The server also supported a real time updated web-interface giving details about the measured agri-parameters (Neelamegam et al., 2007). The closed loop self organizing WSN system used in the study comprised of the following:

- The battery powered nodes with embedded sensors for registering the air temperature and relative humidity were deployed at grid of 30 X 30 m.
- SHT1x is a single chip relative humidity and temperature sensor. The device includes a capacitive polymer sensing element for measuring relative humidity and temperature.
- Networked sensors that measure, and record into an electronics data base, several variables of interest such as soil moisture, soil temperature, pH, ambient relative humidity and ambient temperature. Such automated monitoring system also facilitates the crop experts with a large amount of raw data in electronic formats.
- Each node is able to transmit/receive packets to other nodes inside a well-defined transmission range varying between 30 to 1000 m. A single node can transmit the temperature and relative humidity every minute.
- In a wireless sensor network when the transmission range of a sensor node is not sufficient then it uses multi hop communication to reach the destination node or sink node. For example a node communicates data collected, to a nearby node which in turn transmits to another nearby node in the direction of the sink node. This data forwarding mechanism continues till the sink node is reached. Multi hop communication extends the transmission range of a sensor node and also prevents it from draining soon.
- Signal processing and data processing algorithms that extract useful information out of massive amounts of raw data which is then used to generate alerts that are used to alter sampling frequencies and activate actuators.
- Secure web portal that allows users at different location to access and share their agri-data.
- Solar cell Polycrystalline solar modules (6 V and 500 mA) were used for charging lead acid battery.

2.2.2.1 AgriSens irrigation system

In India, sprinkler and drip irrigation systems are becoming popular irrigation systems. Drip irrigation saves considerable amount of water and hence preferred. As grapevines are arranged in uniform row pattern, drip irrigation is an easy way to control water. Automation can fulfill water requirements of fairly large number of grapevines with single valve. The same pipeline can be used for providing required nutrients also. Grapes are seasonal in India, and they are being planted in month of December to March, which is not a rainy season. Thus, alternate water source has to be used making external source of water essential for grapevines

(Shah et. al., 2009). In vineyard there are different types of grapevines which requires different amount of soil moisture (Burrell et al., 2004). Also, it is very difficult to manually control the irrigation required for particular type of grapevine. WSN based irrigation automation can tackle the problem and also help to save considerable amount of water. The moisture contents of the soil decide the actuator activation. If the threshold level of the soil moisture goes down below a certain level, the valve gets open. This threshold level has to be decided based on climate, topography and type of plant, etc, at the Agri-Information server (Desai et al., 2008). The WSN System, was designed to aid end users and researchers to analyze real time sensor data and assist in decision making for various applications. It was a web based application that could be accessed ubiquitously by the users thus providing a convenient and nimble tool. Since it was integrated with google map, it could provide location-based data. Moreover, this enabled the information to be displayed in a visually readable format.

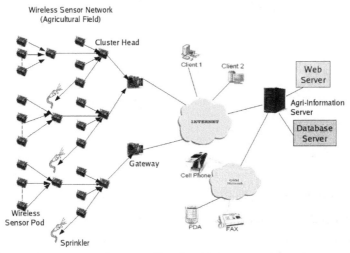

Fig. 2.4. Different Components of Systems Developed at IIT Bombay

2.2.2.2 Sensors suite

Following sensors were deployed based on the feedback received from Sula Vineyard (http://www.sulawines.com) in addition to air temperature and air humidity sensor.

a. Soil moisture sensor

Measuring and monitoring soil moisture helps determining when to irrigate, how much water to apply. The sensor used is ECH2O probe by Decagon as shown in Fig. 2.5 (a). It is a capacitance probe that measures dielectric permeability of medium. In soil, dielectric permeability is related to soil moisture content. Soil moisture was calibrated in terms of volumetric soil moisture content.

b. Soil temperature sensor

The soil temperature shown in Fig 2.5 (b) from Decagon, has a resolution of 0.1∘C. It is enclosed in a low thermal conductivity plastic assembly design to shield the sensor from sunlight and at the same time maximizes convective air movement around the thermistor.

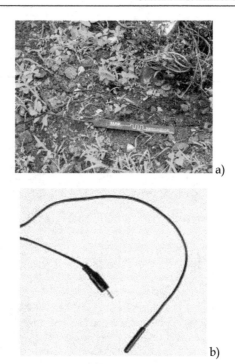

Fig. 2.5. a) Soil Moisture Sensor, b) Soil Temperature Sensor

3. Lessons from the case-study on AgriSens project

3.1 Estimation of Evapo-transpiration (ET) rates for crop okra and grapes

ET is the loss of water from the crop through combined process of soil evaporation and crop transpiration as explained in section 2.1. As discussed earlier, the rate of ET is a function of three critical factors i.e. weather parameters, soil moisture, and nature and stage of growth of the crop. Estimation of ET, to establish the irrigation scheduling using mathematical approach has long been seen as an appealing technique due to simplicity of method when compared with on-site measurements (Allen et al., 1998). ET was estimated using the modified Penman and Moneith model (shown in equation 2.1) to calculate the reference ET and then multiplied with crop coefficient (available in literature) to get the actual crop ET at different growth stages of the crop. The ET for okra was found to vary between 0.1 to 4.0 mm/day, with highest water demand of about 4.0 mm/day during the month of October - December 2007 as shown in Fig 2.7. This is explained by dry climate experienced in Mumbai during month of October to December.

The calculated values of ET for sula vineyard, Nashik were plotted against measured values of soil moisture in Fig. 3.1. Figure 3.1 indicates that soil moisture is influencing the ET loss. This is in agreement with the effect explained by Hatfield and Prueger (2008) and Brown (2000). The rates of ET decreased substantially with decrease in soil moisture content measured over about the top 30 cm depth. Knowing the ideal soil moisture content for crops and given soil texture we can compute the ET and hence irrigation requirement. The

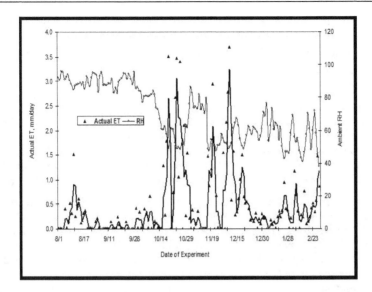

Fig. 3.1. Variation of Evapo-transpiration (ET in mm/day) and Ambient Relative Humidity (RH %) in the Greenhouse, IIT Bombay

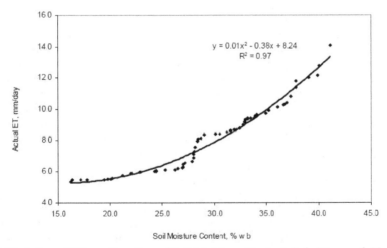

Fig. 3.2. Variation of ET as a Function of Soil Moisture Content, Sula Vineyard, Nashik for the Months of March to May 2008

water requirement through a cycle of 110 days of grape cultivation in the field ranges between 500 to 1200 mm (www.ikisan.com) and the values computed through the sensed parameters in this work ranged from 550-1500 mm. The values of ET for grapes were found to be varying between 5 to 14 mm/day for the months of March till May, 2008. The ET values in grape fields are found to be three times higher than those found in the test bed for okra at IIT Bombay. The field ET for grape crop was computed for the summer months i.e. March to May. The higher ET values for grapevine is further explained by both higher wind velocities in open field and the higher crop coefficients for grapes (0.75) which is almost 1.7 times higher than for okra crop (0.45). The variation in ET values between 5 to 14 mm/day is primarily due to change in soil moisture as the variation in weather data was small.

4. Conclusions

In the past 50 years, world agriculture has experienced enormous changes. Industrialized countries have created a modernized agricultural system with high productivity and advanced technology. Post green-revolution era agriculture in India is facing a technological fatigue for two reasons; a) high rates of ground-water depletion and b) Soil salinity due to excessive irrigation in some pockets. Rapid socio-economic changes in some developing countries are creating new opportunities for application of precision agriculture (PA).

The field deployment case study discussed in the chapter has demonstrated the utility in estimation/saving water use. Weather data monitoring in the shednet house test bed facility at IIT Bombay helped find the ET values for okra ranging between 0.1 to 4 mm/day. The actual ET for grapes in Nashik vineyard, India was found to be varying between 5 to 14 mm/day as the soil moisture varied between 15 to 40 %. While the ET computations were carried out based on data from one season, data for 3- 4 seasons is required for any package of recommended practices as guidelines for entrepreneurs. We believe that WSN supported agriculture management will be particularly useful for larger farms because of its flexibility, more number of sampling points, ease in operation compared to wired sensors- network system. The wide scale appeal of sustainable practices in agriculture and the newer developments in providing low cost/robust sensor based systems are likely to provide the necessary fillip in future agriculture world-wide. Currently the WSN system has high probability of economic viability for high value crops. Despite the widespread promotion and adoption of precision agriculture, the concept of precision irrigation or irrigation as a component of precision agricultural systems is still in its infancy. Some more case studies similar to the one described for other crop-agriculture systems will go a long way in building faith in sensor based irrigation towards both saving precious water as well as soil-degradation due to excessive surface flood irrigation. It also remains to be seen through the field trials that precision irrigation can provide substantially greater benefits than traditional irrigation scheduling. The advances in wireless sensor networks have made some practical deployment possible for various agricultural operations on demonstration scale, which until a few years ago was considered extremely costly or labor intensive. Precision irrigation system with robust components such as, sensing agricultural parameters, identification of sensing location and data gathering, transferring data from crop field to control station for decision making and actuation and control decision based on sensed data will find application in future agriculture. Thus the great potential of integrating the precision farming with WSN to interpolate over a large area for spatial decision making need to be tapped for making agriculture attractive in future.

5. Acknowledgement

The case-study presented here was a part of research project at IIT-Bombay that was supported through a financial contribution from the Department of Information Technology of Ministry of Communications and Information Technology of Government of India. Authors are also grateful for the technical contributions particularly on communication networks (WSN) received from Prof U B Desai and Prof S. N. Merchant of Electrical Engineering Department, IIT-Bombay.

6. References

Adams, M.L.; Cook, S. & Corner, R. (2000). Managing Uncertainty in Site-Specific Management: What is the Best Model? *Precision Agriculture.* 2, 39-54.

Akyildiz, I.F.; Su, W.; Sankarasubramaniam, Y. & Cayirci, E. (2002). Wireless Sensor Networks: A Survey, *Computer Network*, 38, 393-422.

Akyildiz, I.F. & Xudong, W. (2005). A Survey on Wireless Mesh Networks, *IEEE Communication Magazine*, Vol. 43, pp. S23-S30.

Allen, R.G.; Jensen, M.E.; Wright, J.L. & Burman, R.D. (1989). Operational Estimates of Reference Evapo-transpiration, *Agronomy Journal*, 81, pp 650–662.

Allen, R.G.; Pereira, L.S.; Raes, D. & Smith, M. (1998). Crop Evapotranspiration - Guidelines for Computing Crop Water Requirements- FAO Irrigation and Drainage Paper 56. 09-01-2010, Available from http://www.fao.org/docrep/x0490e/x0490e0b.htm.

Booker, J.D.; Bordovsky, J.; Lascano, R.J. & Segarra, E. (2006). Variable Rate Irrigation on Cotton Lint Yield and Fiber Quality. *Belt wide Cotton Conferences*, San Antonio, Texas.

Bronson, K.F.; Booker, J.D.; Bordovsky, J.P.; Keeling, J.W.; Wheeler, T.A.; Boman, R.K.; Parajulee, M.N;, Segarra, E. & Nichols, R.L. (2006) Site-specific Irrigation and Nitrogen Management for Cotton Production in the Southern High Plains. *Agronomy Journal*, 98, 212-219.

Brown, P. (2000). Turf Irrigation Management Series: I, University of Arizona, Cooperative Extension Publication AZ 1194, 09-10-2009, Available from http://ag.arizona.edu/pubs/water/az1194.pdf.

Burrell,J.; Brooke,T. & Beckwith, R. (2004). Vineyard Computing: Sensor Networks in Agricultural Computing, *IEEE Pervasive Computing*, pp 38-45.

Das, Ipsita; Shah N.G. & Merchant, S.N. (2010). AgriSens: *Wireless Sensor Network in Precision Farming: A Case study.* LAP Lambert Academic Publishing, Germany, ISBN: 978-3-8433-5525-4, Germany.

Desai, U.B.; Merchant, S.N.& Shah N.G. (2008). Design and Development of Wireless Sensor Network for Real Time Remote Monitoring. *Unpublished Project Report* submitted to Ministry of Information Technology, Govt of India.

Dukes, M.D. & Scholberg, J.M. (2004) Automated Subsurface Drip Irrigation Based on Soil Moisture. *ASAE Paper No. 052188.*

Evans, R.G. & Sadler, E. (2008) Methods and Technologies to Improve Efficiency of Water Use. *Water Resources Research*, 44.

Goodwin, I. & O'Connell , M.G. (2008). The Future of Irrigated Production Horticulture – World and Australian Perspective. *Acta Horticulturae*, 792, 449–458.

Hatfield, J.L. & Prueger, J.H. (2008). Encyclopaedia of agricultural, food, and biological engineering. D. R. Heldman (Ed). New York: Marcell Dekkar, pp 278-281.

Hedley, C.B. & Yule, I.J. (2009). Soil Water Status Mapping and Two Variable-Rate Irrigation Scenarios. *Precision Agriculture*, 10, 342-355.

Jensen, M.J. (1981). Summary and Challenges. *In Proceeding of the ASAE's Irrigation Scheduling Conf., Irrigation Scheduling for Water & Energy Conservation*, American Society of Agricultural Engineer, St. Joseph, MI. pp. 225-231.

King, B.A.; Stark, J.C. & Wall, R.W. (2006). Comparison of Site-Specific and Conventional Uniform Irrigation Management for Potatoes. *Applied Engineering in Agriculture*, 22(5), 677-688.

Ladha, J. K.; Fischer, A. K.; Hossain, M.; Hobbs, P. R. & Hardy, B. (2000). Improving the Productivity and Sustainability of Rice-Wheat Systems of the Indo-Gangetic

Plains: IRRI Discussion Paper Series No. 40. Makati City (Philippines): International Rice Research Institute. pp 31.

McBratney, A.B.; Whelan, B.; Ancev, T. & Bouma, J. (2005). Future Directions of Precision Agriculture. *Precision Agriculture*, 6, 7-23.

Modak, C.S. (2009). Unpublished Report from Land and Water Management Institute, Aurangabad, Maharashtra, India.

Mondal, P. & Tewari, V.K. (2007). Present Status of Precision Farming: A Review, *International Journal Agricultural Research*. 2(1), pp 1–10.

Neelamegam, S.; Naveen, C. P. R. G.; Padmawar, M.; Desai, U. B.; Merchant, S.N. & Kulkarni, V. (2007). AgriSens: Wireless Sensor Networks for Agriculture- Sula Vineyard A Case Study, *1st International Workshop on Wireless Sensor Network Deployments*, Santa Fe, New Mexico, U.S.A. June 18 – 20.

Panchard, J.; Rao, S.; Prabhakar, T. V.; Hubaux, J.P. & Jamadagni, H. (2007). Common Sense Net: A Wireless Sensor Network for Resource-Poor Agriculture in the Semi-arid Areas of Developing Countries. *Information Technologies   and International Development*, Volume 4, Number 1, 51–67.

Pierce, F.J. & Nowak, P. (1999). Aspects of Precision Agriculture. *Advances in Agronomy*, 67, 1-85.

Roy, S.; Anurag, D. & Bandyopadhyay, S. (2008). Agro-sense: Precision Agriculture using Sensor-Based Wireless Mesh Networks. Innovations in NGN: Future Network and Services, In: *Proceedings of the First ITU-T Kaleidoscope Academic Conference*, Geneva, Italy, pp 383–388.

Sadler, E.J.; Evans, R.G.; Stone, K.C. & Camp, C.R. (2005). Opportunities for Conservation with Precision Irrigation. *Journal of Soil and Water Conservation*, 60(6), 371-379.

Schoengold, K.; Sunding, D.L. & Moreno, G. Agricultural Water Demand and the Gains from Precision Irrigation Technology. 12-08-2011, Online available from http://giannini.ucop.edu/media/are-update/files/articles/v7n5_2.pdf.

Shah, N.G.; Desai U.B.; Das, I.; Merchant, S.N. & Yadav, S.S. (2009). "In-Field Wireless Sensor Network (WSN) For Calculating Evapo-transpiration and Leaf Wetness, *International Agricultural Engineering Journal*, 18(3-4), pp 43- 51.

Smith, R.J.; Raine, S.R.; McCarthy, A.C. & Hancock, N.H. (2009). Managing Spatial and Temporal Variation in Irrigated Agriculture Through Adaptive Control. *Australian Journal of Multi-disciplinary Engineering*, 7(1), 79-90.

Stafford, J.V. (2000). Implementing Precision Agriculture in the 21st century. *Journal of Agricultural Engineering Research*, 76, pp 267-275.

Wang, M. (1998). Development of Precision Agriculture and Engineering Technology Innovation, *Transactions of the CSAE, Chinese Society of Agricultural Engineering*. Beijing, 15(1), pp 231.

Wang, N.; Zhang, N. & Wang, M. (2002). Precision Agriculture, A worldwide Overview, *Computers and Electronics in Agriculture*, 36, pp 113-132.

Wang, N.; Zhang, N. & Wang, M. (2006). Wireless Sensors in Agriculture and Food Industry — Recent Development and Future Perspective, Review: *Computers and Electronics in Agriculture*, 50, 1–14.

Whelan, B.M.; McBratney, A.B. & Boydell, B.C. (1997). The Impact of Precision Agriculture. *Proceedings of the ABARE Outlook Conference*, Moree, UK, July, pp 5-7.

Yule, I.J., Hedley, C.B. and Bradbury, S. (2008). Variable-rate irrigation. 12th Annual Symposium on Precision Agriculture Research & Application in Australasia. Sydney.

4

Criteria for Evaluation of Agricultural Land Suitability for Irrigation in Osijek County Croatia

Lidija Tadić
Faculty of Civil Engineering
University of J. J. Strossmayer Osijek
Croatia

1. Introduction

Considering the basic purpose of agriculture – ensuring of sufficient quantities of food with appropriate quality and unquestionable health soundness, the management of land should not sideline   environmental and social aspects. Today, modern agriculture and rural development besides food production, involves ecosystems sustainability and rising values of landscape and rural space in general. Irrigation also must be part of this approach to modern food production. Irrigation in Europe is developing continuously, especially in Mediterranean countries (France, Greece, Italy, Portugal and Spain) resulting in the increase of water quantities used for irrigation. Many countries which recently joined the EU also develop their irrigation systems and this trend will probably continue. So, the future development must be done according to local natural conditions, social and economic background. As an undertaking which ensures optimal water supply for demands of agricultural production, the irrigation might have significant impacts on environment, which should be foreseen and targeted by use of economically acceptable activities in order to eliminate or lower those potentially negative impacts to the acceptable levels. The aim of every project of the hydro-melioration system, including irrigation is to ensure positive long-term effects of the implemented system which is achieved by: anticipation of potential problems, defining the means of monitoring, finding the ways to avoid or reduce problems and promotion of positive effects (Tadić, Bašić, 2007).

It can be said that importance of irrigation grows every year, even in the countries which are not located in arid or semi-arid regions. General objectives of irrigation implementation in any given area are:

- Increasing of agricultural production and stability of production during dry years,
- Introduction of new more profitable crops on the market,
- Reduction of food import and stimulation of domestic agricultural production,
- Reduction of climate change impacts, first of all frequent drought periods,
- Reduction of agricultural land,
- Negative water balance during the vegetation period,
- Increase the interest for farming and employment in the agriculture (Romić, Marušić, 2005)

Irrigation systems should be based upon principles of integrated water resources management and sustainable management taking under consideration potentials and restrictions of specific river basin. Several levels of data evaluation are needed:

- **Physical plans give basic information of agricultural areas, present state and future development, possible increase or decrease of the area.**
- **Soil properties are one of the most important factors because soil categorization from very suitable to not suitable for irrigation has a great impact on the final decision.**
- **Agricultural potential, mechanisation and other resources of agricultural production, including tradition of growing crops and interest of local people for irrigation. Part of the agricultural potential is developed land drainage and flood protection system.**
- **Analysis of hydrological and meteorological data, particularly analysis of drought, defines the real necessity for implementation of irrigation. Frequency, intensity and duration of water deficit in the vegetation period indicate the crop water requirement which has to be assured.**
- **Availability of water resources is one of the most restrictive criteria, where two aspects have to be considered – water quantity and quality. In the area with evident water shortage, for example Mediterranean islands, some alternative sources of water have to be applied.**
- **All possible environmental impacts of irrigation implementation should be recognized in the area due to its vulnerability and sensitivity on changes by its structure or genesis (e.g. karst regions).**

Basically, above mentioned levels of evaluation are given according to the DPSIR (driving force- pressure-state-impact-response) relationship, which is dynamic in time. Sustainability of the irrigation system can be achieved only by its constant improvement and development. Figure 1 gives a scheme of sustainable use of water resources in irrigation.

If any of the components in the DPSIR relationship changes in time, the whole scheme changes as well. The aim of sustainable approach is to achieve positive movement of the whole process, which means the decrease of pressure and unwanted impacts together with improvement of the state with ensuring domination of positive impacts. This process will be possible with strong response development.

Starting point in irrigation development is the initiative of farmers, investors and final users of the irrigation system. All other phases will be elaborated in the following chapters.

2. Physical planning

The irrigation systems are very pricey and complex undertakings and their implementation needs clear economic analysis, and no omissions should be allowed during their planning.

The initial phase in planning of irrigation systems is identification of spatial limitations as defined in physical planning documentation of either Municipality or County. Physical planning documentation, apart from natural and social characteristics of the analysed area, define the scope of economic development, including transportation, electric power, water management and other activities within the space, as well as limitations to construction in regard to protected areas.

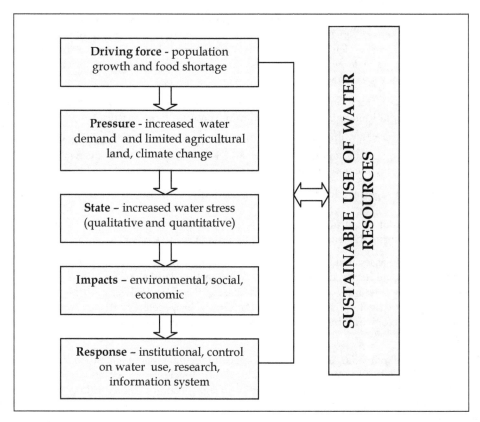

Fig. 1. Scheme of sustainable approach to irrigation (Boss, Burton 2005)

Anticipated change in landuse of a certain area, i.e. from agricultural to construction land or intersecting of the agricultural area with the large-scale infrastructure project such as road or railway, derivative channel, navigable channel or transmission line may impact a decision regarding withdrawal of construction of the irrigation system on certain agricultural land regardless of needs and natural potential.

Some of the proposed activities in space do not need to fully stop development of an irrigation system but may significantly influence the choice of irrigation methodology. These may also provoke additional costs towards implementation of environmental protection measures in the case if agricultural land is located within protected water well site or in the vicinity of areas protected for their natural values. Using physical planning documents it is possible to establish availability of water resources as well as planned activities on waterways.

For determination of agricultural land suitability for irrigation and economic feasibility of the system it is very important to consider the existing access road and electric power infrastructure, which are very important aspects for the investment project.

3. Soil suitability for irrigation

The following group of parameters which define the need for irrigation but also the its' methodology are characteristics of soil. The soil characteristics are the result of paedogenesis factors and processes, which have been influencing their mechanical, physical and chemical properties during the long period of time, and the most importantly its capability of movement and storage of water in the soil.

Along with these factors long periods of cultivation and use in intensive agriculture have contributed to significant anthropogenesis of initial natural characteristics.

Soil in this case is being analysed as a cultivation space in which during the vegetative period is lacking specific amount of water and which has to be introduced by artificial means, while getting the best possible effects:

• Maximum usage of added water which is firstly, economic but also environmental condition;
• Preservation of soil structure – during which it is necessary to assess the water regime in the soil during the entire year, especially if there is a risk of soil salinisation.

Therefore, for the assessment of soil potential for irrigation the following parameters are of key importance: soil depth, drainage and flood protection, land slope and erosion potential, water capacity, soil salinity, quantity of nutrients, etc. Based on analysis of these parameters, soli potential for irrigation is evaluated for the area of proposed activity, and also the most suitable methodology and measures required for improvement of existing soil potential are proposed.

Considering the potential for irrigation, soils are usually being classified as excellent (P-1), suitable (P-2) and restrictively suitable (P-3). In classes of potential P-2 and P-3 it is necessary to undertake hydro-technical and agro-technical interventions of different scope, which would increase the degree of their potential for irrigation (Tadić, Ožanić, et.al.2007).

This creates additional costs which investor has to bear in order to implement proposed irrigation system, but this system will then enable rational usage of water, preservation of soil characteristics and limiting of unwanted consequences to a minimum.

In specific cases, there is an occurrence of soil characteristics, which would categorise the soil as permanently inadequate for implementation of irrigation because the costs of soil characteristic's improvement would be immensely high and would impact the economic feasibility of the system, i.e. areas without surface and ground drainage, which is necessary due to retention of water in plant roots zone, or areas with a significant inclination which are prone to water erosion.

For example, Figure 2 presents total agricultural area of Osijek County, which is about 280.000 ha (100%). Part of it belongs to the Nature Park and well site for public water supply, and intensive agricultural production are not allowed (30%). Part of the area belongs to the roads, water bodies and settlements (33%), and about 1% is completely unsuitable for irrigation. The rest of the agricultural land, about 36%, can be considered as more or less suitable for irrigation according to physical and chemical properties, environmental characteristics and drainage conditions. This type of soil evaluation was made in order to avoid negative irrigation effects on the soil and to minimize the costs of land reclamation.

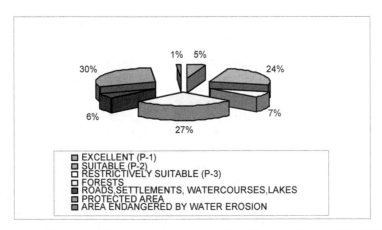

Fig. 2. Example of land suitability for irrigation in Osijek County , Croatia (Tadić, 2008)

4. Agricultural potentials

Introduction of the irrigation system implies the existence of interest of agricultural producers for improvement in production means of agricultural goods. The aim of irrigation is increase and stabilisation of yield and increase in market value of final product. Financial investments in irrigation are considerable, which means that the users should have on their disposal land, machinery, but also the knowledge which would enable them to reach the wanted goal. Existing land protection from outside waters (flood protection) and excessive surface and ground waters, and existence of the drainage system for surface and ground waters are important prerequisites for construction of an irrigation system. Agricultural potential also means readiness for introduction of new crops, which are more profitable on the market and which cultivation is impossible without irrigation.

In Croatia, majority of the continental part of the country is the traditionally agricultural area with relatively high level of agricultural production with regard to land organization, as well as production technology, mechanization and application of scientific and contemporary methods in agricultural production. Main characteristics of present crop production are as follows: orientation toward an extensive food production and small number of cultivated crops that are present on small acreage, low presence of fruit and vegetables and inadequate use of quality land resources suitable for production diversification (Tadić et al. 2007). Introduction of irrigation could enlarge the strength of present agricultural production with the introduction of more profitable, water dependent sensitive crops.

5. Irrigation necessity

The most common way of elaboration of irrigation necessity is an analysis of water deficit during the vegetation period. It causes reduced actual evapotranspiration compared to the potential evapotranspiration. Water deficit (drought) in different crop development stages causes stagnation in growth and finally reduces the yield. Drought damages depend on duration, strength and intensity of drought, which is basically characterized by geographical characteristics, soil type and sort of crop as well. In Croatia, the most severe damages are in vegetable yields, and the least in cereal yields.

During recent years many results were published, which prove damages caused by drought during the vegetation period. Table 1 presents yield reduction for crops, which are specific in Croatia.

The emphasis is given to the influence of soil type. Light soil with poor water retention capacity is more vulnerable to the water deficit.

CROP	Water deficit (mm)		Yield reduction (%)			
	Average year	Dry year	Light soil		Heavy soil	
			Average year	Dry year	Average year	Dry year
Corn	187	287	35,4	61,5	27,1	54,4
Sugar beat	246	349	36,7	58,7	29,1	52,6
Tomato	188	260	37,3	54,8	29,8	47,9
Apple without mulch	180	294	25,7	49,4	20,8	41,8

Table 1. Water deficit and yield reduction in average and dry years observed depending on soil conditions (Tadić,2008)

The fact that different crops have different sensitivity to water deficit affects the decision on irrigation implementation and cost-benefit analysis of a planned irrigation system. Intensity of crop diversification and implementation of irrigation are strongly influenced by trends on the market.

Decrease of the yields due to the water deficit can be expressed in several ways. One of them is the linear statistical relationship between total evapotranspiration and yield of cereal grains in some climate zone (Hoffman et al.2007). It can be given by equation:

$$Y_g = bET + a \tag{1}$$

where: Y_g= yield (t/ha)
         ET= vegetation season evapotranspiration (mm)
         b=slope of the yield-ET line (t/ha mm)
         a=constant (t/ha).

The second one is very often used relation proposed by FAO (Doorenbos et al.1986):

$$\left(1 - \frac{Y_a}{Y_m}\right) = k_y\left(1 - \frac{ET_a}{ET_m}\right) \tag{2}$$

where: Y_a = actual harvested yield (t/ha)
         Y_m= maximum harvested yield (t/ha)

k_y= yield response factor
ET_a= actual evapotanspiration (mm)
ET_m= maximum evapotranspiration (mm)

Increase in yield can be achieved by the sufficient amount of available water which will increase actual evapotranspiration in the vegetation period during crucial crop development stages.

6. Drought analysis

Draught ,may vary in time and space, depending on climate and hydrological conditions of some area. According to World Meteorological Organization, drought is a protracted period of deficient precipitation with high impacts on agriculture and water resources. There are three types of droughts: meteorological, agricultural and hydrological drought. Any of these types can make a serious harm on agricultural production and economy in general. Long and frequent drought periods can cause desertification characterized by water shortage, overexploitation of available water resources, change and depletion of natural vegetation, reduction of crop varieties, reduction of water infiltration, etc. (EEA, 2000). Proper analysis of drought as an extreme hydrological event is essential for identification of irrigation necessity as a successful measure against it. In another hand, the same level of drought severity can cause different impacts in different regions due to the underlying vulnerabilities. Basic meteorological and hydrological data, precipitation, air temperature, evapotranspiration, relative humidity, wind, insolation and discharges must be available to provide proper analysis of drought. Figure 3 presents total annual precipitation and precipitation in growing period observed in the period from 1951 to 2000 on meteorological station Osijek, Croatia. The both trend lines show decreasing of precipitation. There is no significant decreasing during the vegetation period, which is good, but it indicates smaller possibilities of groundwater recharge during winter time.

Figure 4 presents average annual air temperature, and average air temperature during growing season observed also in the period from 1951 to 2000 on meteorological station Osijek. Increase of air temperature is obvious in both figures, but it is still lesser during the vegetation period.

There are a lot of methods for drought estimation, developed for different climates and geographical features. Analysis of drought is very complex and the identification of a moment when drought starts and becomes extreme is very sensitive and variable.

There are two basic groups of methods: agro-climatic and hydrological. The methods in the first group are mostly based upon above mentioned two parameters, precipitation and air temperature. They give characteristics of the climate in some area with regards to the agriculture. The most common are:

• Lang's precipitation factor (KF)

$$KF = \frac{P}{T} \qquad (3)$$

where: P= annual precipitation (mm) and T=average annual air temperature (ºC)

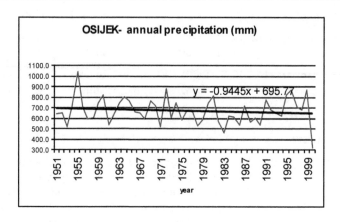

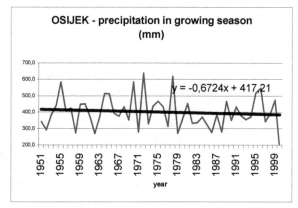

Fig. 3. Average annual precipitation and precipitation in growing season in the period 1951-2000, Osijek, Croatia

- Thronthwaite's humidity index (I_{pm})

$$I_{pm} = 1,65\left(\frac{P}{T + 12,2}\right)^{10/9} \tag{4}$$

where: : P= annual precipitation (mm) and T=average annual air temperature (°C)

- de Martone's draught index (I_s)

$$I_s = \frac{P}{T + 10} \tag{5}$$

where: P= annual precipitation (mm) and T=average annual air temperature (°C)

- Walter's climate diagram gives relationship between precipitation and air temperature or evapotranspiration and in that way indicates the necessity of irrigation and the length of irrigation period . According to Walter's climate diagram, irrigation is necessary only in July and August .

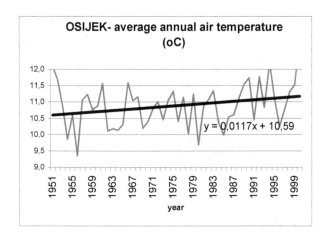

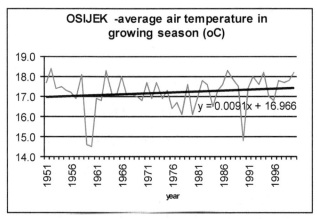

Fig. 4. Average annual air temperature and air temperature in growing season in the period 1951-2000, Osijek, Croatia

Table 2 presents climate description obtained by three of these methods applied on data of precipitation and air temperature for Osijek (1951-2000).

Lang's precipitation factor (KF)		Thronthwaite's humidity index (I_{pm})		de Martone's draught index (I_s)	
Extremes and average values	Climate description	Extremes and average values	Climate description	Extremes and average values	Climate description
$KF_{min}=24,6$	Arid	$I_{pm\,min}=20,8$	Semi-arid	$I_{s\,min}=13,9$	Dry sub-humid
$KF_{aver}=62,2$	Sub-humid	$I_{pm\,aver}=8,1$	Sub-humid	$I_{s\,aver}=32,3$	Humid
$KF_{max}=98,4$	Humid	$I_{pm\,max}=75,4$	Humid	$I_{s\,max}=50,6$	Per-humid

Table 2. Drought estimation obtained by Lang's precipitation factor, Thronthwaite's humidity index and de Martone's drought index

Results in Table 2 indicate climate characteristics of the region, considering annual data. They vary in the range from arid to per-humid. The average values show sub-humid to humid climate characteristics. Besides, Walter's climate diagram shows the need of irrigation in June and July what corresponds with other approaches.

The methods in the second group also use meteorological and hydrological parameters for drought identification, but they can be considered as more comprehensive and reliable.

Because of the great variety of approaches and methods, it is recommended to use more than one method to estimate an intensity of drought periods and necessity of irrigation. In that way, it is possible to give more accurate estimation of water deficit in some region. Some of the most frequently used methods will be briefly explained.

- Standard Precipitation Index (SPI) is one of the most popular methods, proposed as a most appropriate method for any time scale and any region in the world. The SPI is an index based on the probability of precipitation using the long-term precipitation record. A drought event begins when the SPI is continuously negative and ends when SPI becomes positive (WMO).

$$SPI_n = \frac{1}{\sigma_n}\left(\sum_{i=1}^{n} P_i - P_n\right) \qquad (6)$$

where: n= number of monthly precipitation data,
         P_i = precipitation in each month (mm),
         Pn= average precipitation of the observed period (mm),
         σ_n =standard deviation.

- Deciles is a method in which monthly precipitation sums from a long term record are first ranked from highest to lowest to construct a cumulative frequency distribution. The distribution is then divided into 10 parts (deciles). A longer precipitation record (30-50 years) is required for this approach.
- The Rainfall Anomaly Index (RAI) also ranks the precipitation data of the long-term period in the descending order. The average of the 10 highest values as well as that of the 10 lowest precipitation values were calculated (Figure 7).

$$RAI = \pm 3\frac{P - \overline{P}}{\overline{E} - \overline{P}} \qquad (7)$$

where $\overline{P}$ is average of the annual precipitation for each year (mm) and $\overline{E}$ is average of 10-extrema for both positive and negative anomalies (mm).

- The Stochastic Component Time Series (SCTS) is given by the equation

$$Z\varepsilon = \frac{\varepsilon t - \overline{\varepsilon}}{\sigma \varepsilon} \qquad (8)$$

where εt is total annual rainfall for each year (mm), $\overline{\varepsilon}$ is average annual rainfall for each year (mm) and $\sigma \varepsilon$ is standard deviation of rainfall for each year (Figure 5).

These methods applied on precipitation data observed on Osijek meteorological station (1951-2000) give results in Figure 5 and Table 3.

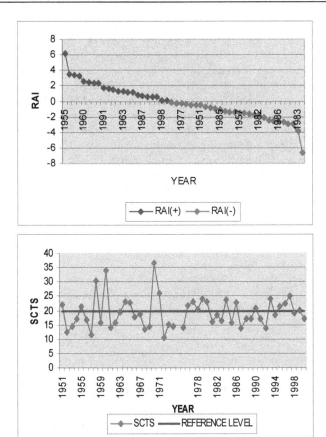

Fig. 5. Application of RAI method and SCTS method on precipitation data of Osijek, Croatia (1951-2000)

Rainfall Anomaly Index (RAI) shows that 28 years in the observed period were dry and 22 were less dry. According to Stochastic Component Time Series (SCTS) 29 years were dry and 21 years less dry.

Standard Precipitation Index (SPI)		Deciles	
Value	Classification of dryness	Months (%)	Classification dryness
		23	Average
-4,4	Extremely dry	56,5	Below the average
		20,5	Very much below the average

Table 3. Application of SPI Method on precipitation data of Osijek, Croatia (1951-2000)

Standard Precipitation Index (SPI) indicates the considered period extremely dry and Deciles classified 77% of the total number of months in the 50 years long period as below and very much below the average.

In the hydrological analysis, it is very often stressed that data series of precipitation, discharge, water level, etc., should be as long as it is possible. The length of the data series should guarantee reliability of final results and conclusions. This can be questioned due to the process present in recent decades, often referred to as a climate change. Hydrological analysis based upon a long-term data series (50, 70 or 100 years) sometimes can lead to the false conclusions. In that case, it is recommended to apply the RAPS (Rescaled Adjusted Partial Sums) method. This analysis helps us to recognize any change in time series, like periodicity, sudden leaps or smaller errors in the data series. Figure 8 and 9 present two examples of RAPS implementation. The example on Figure 6 shows annual precipitation data (1948-2008) of Vela Luka and Korčula, small places in Korčula Island in Adriatic Sea. The complete data series show the negative trend line of the same magnitude. Application of RAPS resulted by obvious break in the both data series, which occurs in 1982 and the both sub-series have the positive trend line (Ljubenkov, 2010).

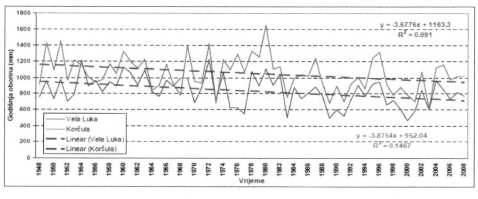

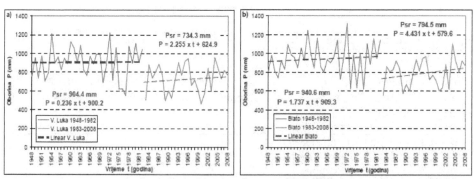

Fig. 6. Application of RAPS method on precipitation data of Korčula, Croatia (1948-2008) (Ljubenkov, 2011)

Figure 7 presents application of RAPS method on annual precipitation data of Osijek (1951-2000) and it shows two breaks, in 1974 and 1990. The first data sub-series 1951-1974 have a mild positive trend line, and next two sub-series have emphasised negative trend lines. Comparing to the complete annual data serine presented in Figure 3 with the continuous negative trend line the difference is significant.

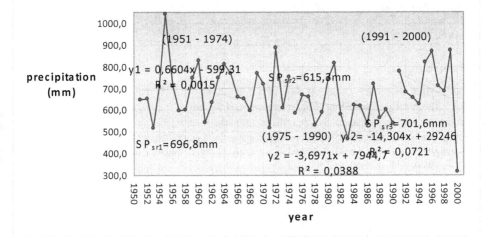

Fig. 7. Application of RAPS method on precipitation data of Osijek, Croatia (1951-2000)

7. Water availability

The next important precondition for implementation of irrigation systems is availability of water and is one of the most limiting factors. Water resources are estimated in accordance to three criteria: quantity, quality and location. Together they form water potential (Đorđević, 1990). Quantities of water required for irrigation are considerable and depend on water deficit, crops and size of the area. All three characteristics have to be met in order for water capture to satisfy long-term needs and if only one of them is not met, quantity, quality or accessibility of location such source of water becomes questionable. Usual sources of water for irrigation are rivers, ground water tables and reservoirs, and in the more recent times various non-conventional sources. In the majority of Europe and world, the shortage of water is becoming more emphasised, especially in some regions, and every new consumer of water significantly affects the water balance. Large rivers often have multiple uses: navigable waterway, electric power generation, capturing of water for different purposes, the recipient of waste waters and provision of quality water environment for a number of species, and all these requirements have to be met without conflicts in interest. About 30% of abstracted water in Europe is used for irrigation, but in some Mediterranean countries (Italy, Spain, Greece) this percentage exceeds 55-80% (Nixon et al, 2000).

7.1 Open watercourses

Favourable conditions found in open water bodies for capturing of water for irrigation depend on their hydrological regime. Waterways with the glacial hydrologic regime are more suitable, since they have maximum flows during beginning of summer (June), while waterways with the pluvial regime have minimal flows during summer months in vegetation periods. It is important not to forget the global trends in the decrease of water quantities, increase of frequency and duration of low water regimes, which also cause droughts (Bonacci, 2003). Therefore, capturing possibilities are significantly reduced at low and average water flows for purposes of irrigation during vegetation periods. The limiting factor for such waterways is maintenance of biological minimum required for sustaining of life in water. Quality of water in open waterways generally meets requirements for irrigation.

7.2 Ground water

Ground water usage in irrigation is a very sensitive, but also easily available and cheap solution. Its quality is excellent and there are no qualitative restrictions for irrigation or any other usage. In the most of the countries it is the major public water supply resource. Groundwater overexploitation can endanger capacities of public water supply systems. That is the basic reason of its protection by heavy regulations in many European countries, and in Croatia as well.

Ground water exploitation is especially sensitive in the coastal region due to the possibility of salt water intrusion in aquifers.

In some countries, like Germany, Portugal, Lithuania special permission for groundwater usage in irrigation is required, and the exploited quantities are supervised by official persons (Tadić, Marušić, 2008). In Croatia, possibilities of groundwater usage in irrigation are limited to the resources which can be renewed in the one year.

7.3 Reservoirs

The construction of reservoirs represents a very acceptable but expensive solution applicable in cases where no other sources of water are present. Since reservoirs are very expensive structures they are often constructed as multipurpose structures having to meet requirements of all users, which may lead to conflict of interest, for example, generation of electric power and usage of water for irrigation. Quality of water in the reservoir depends on its geographic location and surrounding area, but in most cases meets the required quality for irrigation.

Natural lakes can have a great potential for usage in irrigation in the quantitative and qualitative sense. However, they are very often protected due to their biological values and landscape features.

7.4 Non-conventional water resources

Use of non-conventional water resources imply the use of treated waste water, rainwater, saline water and melted snow in regions where there are no sufficient amounts of the water present. Use of these sources of water for irrigation in a safe way for agricultural land and environment altogether, is very expensive, and is made possible by intensive development of technology for water treatment. Waste water re-use and seawater desalination are increasing in Europe (e.g. Southern Europe). Application of re-used water should be subject to more research on health aspects.

In Croatia due to wealth of water resources, these forms of water capture are rare, but it is recommended to use captured rainwater on Croatian Islands (Bonacci, 2003).

8. Environmental issues

In the case where there are protected areas in the vicinity of agricultural land, implementation of irrigation should not make any negative impacts on them. All environmental impacts of irrigation should be recognized and evaluated. According to the categorisation of environmental impacts, the expected impacts, which arise due to application of irrigation are:

- According to the *type of impact* – impacts on natural assets, predominantly on water and soil (physical environment), but also on quality of life (social-economic impacts),
- According to the *duration of impact* – long term,
- According to *the occurrence in time and space* - direct, since they occur on exact area, which is being irrigated and during the period of irrigation, but also *indirect*, which means that they also impact the downstream and upstream soils, and frequently appear only after significant periods of time,
- According to the *number of impacts* - individual and cumulative (Tadić, 2009).

All those elements make the impacts very complex and hardly predictable, while the intensity of their occurrence depends on properties of the watershed, water abundance, properties of soils, quality of water being used for irrigation, as well as depending on the applied methodology and means of irrigation.

This implies that the application of irrigation may leave permanent (irreversible) consequences on the environment if the impacts are not recognised, foreseen and possibly mitigated or completely prevented. Some of the changes are easily noticed and quantified, but there is a vast number of indirect impacts that are delayed in time after the prolonged application of irrigation and often appearing outside of the irrigated area. The solutions are found in systematic planning, designing, construction and operation of undertaking. For this reason, large-scale irrigation projects should include environmental impact assessment prior to the construction which will establish the possible alteration to the environment and assess the sustainability of the system.

8.1 Impacts on water

The irrigation has quantitative and qualitative impacts on surface and ground waters.

8.1.1 Impacts on water balance

Any capture of water will impact the existing water balance. Considering the occurrence of water resources in time, every uncontrolled capture, especially in dry periods, may result in undermining of minimum biological requirements of waterways. Some watercourses have minimum flows at the time of vegetation growth when there is a need for irrigation. In smaller waterways and streams this issue is even more pronounced. Hydrologic regime of surface waters is directly related to the levels of ground waters (Romić, Marušić, 2005). During the dry periods, ground waters feed the waterways while in the period of high-water levels, surface waters feed the ground waters. Intensive capturing of surface waters combined with usually water level slope result in increased hydraulic gradient in relation to ground waters. Impacts of capturing of water above renewable limits may appear after prolonged periods of utilisation and may result in lowering of ground water levels on wider area. In coastal areas lowered levels of ground waters may cause intrusion of salt water. Continuous lowering of ground waters, along with changes in water balance, may have effect on other economic activities and water customers. Such changes have the significant impact on sensitive ecosystems, firstly, on low-lying forests and wetlands.

As it was mentioned before, one of the solutions for ensuring supply of sufficient quantities of water for irrigation is construction of water reservoirs. Such structures are considered to be very sensitive hydro-technical undertakings, especially if they are reservoirs with large volume and area, which may have significant impacts on the environment, including both positive and negative effects. With construction of reservoirs, there is a change in land-use of the area (Tadić, Marušić, 2001). The reservoirs are considered as the most controversial hydrothechnical structures which on the one side enable rational use of water collected during wet periods of the year (flood protection), and from the other side have number of environmental consequences – change of landuse, impacts on landscape and wider environment, change of hydrological and biological characteristics of a waterway. Land area is being turned over into a water surface, which changes the fundamental biological structure. Furthermore, the transition from natural to the controlled regime of a waterway after construction of the reservoir causes the number of changes. One of them is a reduction of sediment transport, which is being accumulated and deposited within the reservoir along with increased kinetic energy of water, which affects river bed and banks downstream. Reservoirs have positive effects on the regime of low and high water level periods and consequently, on replenishing of ground water resources in the downstream area.

Changes in the hydrologic regime related to capturing of water may increase concentration of water pollution and generally affect the good status of water quality. Areas exceptionally sensitive to changes in water balance are protected ecosystems whose subsistence is dependent on sufficient water quantities, water capture areas, waterways with decreasing characteristic of water flow trends and coastal areas. One of the sensitive water resources are narural lakes and use of water from natural lakes is not recommended. Some of the lakes in Croatia are already under protection, and there is an incentive to protect all natural lakes in order to preserve values of their ecosystems (Romić, Marušić, 2005).

8.1.2 Impacts on water quality

Water pollution is broad term, but it is generally defined as the reduction of quality due to introduction of impurities and potentially harmful substances. Agriculture is one of the largest non-point sources of water pollution, which is generally hard to identify, measure and monitor. The irrigation is undertaking, which impacts the changes in the water regime of soil, and consequently, on transport of potentially harmful substances to the surface and ground waters (nitrates, phosphorus) causing the eutrophication. Plant manure, residuals of pesticides and other components of agricultural chemicals in natural and irrigated conditions with changed water balance are subject to flushing from soil, and as such they represent a pollution threat to water resources. The speed and intensity of pollution transport from soil depend on a number of factors related to hydrogeological and soil characteristics of the area. In this regard, the especially sensitive are karst and alluvial areas with the relatively thin topsoil layers. Possible protection measures include:

- Adjustment of existing regulations to international standards, or regulation of issues, which are not so far covered by the laws (Ayers, Westcot, 1985)
- Setting up of monitoring system, especially in case where irrigation is present;
- Setting up of an efficient supervision system.

8.1.3 Protected areas

Significant limitations to intensification of agriculture also referring to irrigation are areas under protection. For example, protected drinking water areas in Republic of Croatia amount to 19 % of land areas, while regulations are limiting agricultural production within zones I and II of sanitary protection, with zones III and IV of sanitary protection having no limitations. Meanwhile on water protection areas there should be no priority development of irrigation projects, because of protection applied to water resources aimed at drinking water supply. However, currently there are 2200 km² of protected areas used for agricultural production, with different types and intensity of utilisation (Romić, Marušić, 2005). In the case that within protected areas, and in compliance with valid regulations, there is a justified plan for intensive use of land for agriculture and construction of the irrigation system, it is required to complete the environmental impact assessment which will provide answers if the proposed technology of agriculture may have significant negative impacts on protected component of environment or on any other component of ecosystem. Possible protection measures may include:

- Controlled capture of surface water along with preservation of biological minimum and other requirements (water supply, hydro-energy, inland navigation),
- Controlled capture of ground waters within renewable limits,
- Ensuring of biological minimum in waterways on which reservoirs are built,
- The preference is given to smaller reservoirs over bigger ones,
- Discharge of sediment from reservoirs for safeguarding equilibrium within the waterway,
- Monitoring of ground water levels on wider area of undertaking,
- Monitoring of low water flow trends.

8.1.4 Impacts on biosphere

Changes in land-use of area and changes within ecosystems for purposes of agricultural production, along with application of irrigation, have direct impacts on biosphere. Transition of non-fertile land with specific ecosystems developed (wetland, forest and meadow ecosystems with great biological diversity), which was the common practice not so long ago is now forbidden and not practised any more.

Secondary or indirect impacts on biosphere as a consequence of irrigation may appear with significant reduction of ground water levels, which impairs biological conditions within an ecosystem. According to the Croatian Law (*Law on environmental protection*, OG 82/94) the main aims of environmental protection are permanent preservation of biological diversity of natural communities and preservation of ecological stability, followed by preservation of quality of living and non-living environment and rational use of natural resources, preservation and regeneration of cultural and aesthetic values of landscape and improvement of environmental state and safeguarding better living conditions (Tadić, 2001).

9. Sustainable irrigation

Previously elaborated matter deals with the main characteristics of agricultural land, which should be analysed and evaluated in order to define land suitability for irrigation. The main objective of land evaluation for irrigated agriculture is to define actual physical needs for irrigation and to predict future conditions after development has taken place. Therefore, all relevant land characteristics, including soil, climate, topography, water resources, existing and planned agricultural production, etc. and also socio-economic conditions and infrastructure need to be considered. Some factors that affect land suitability are permanent, and others are changeable at a cost. Typical examples of almost permanent features are meteorological characteristics, basic soil characteristics, topography and landscape. Features changeable with costs are water resources, agricultural potential and soil suitability for irrigation. The costs of necessary improvements can be determined (e.g. construction of a drainage system), so the economic and environmental consequences of development can be predicted. Figure 8 gives a scheme of evaluation criteria (A) and their influence on irrigation suitability (B) in large-scale projects. The third part of the relation (C) is more related to the projects of smaller scale. Parameters of irrigation methods and performance (C) must have separated economic analysis, more detailed and adjusted to the specific project. Five groups of criteria (A) can significantly reduce total agricultural area.

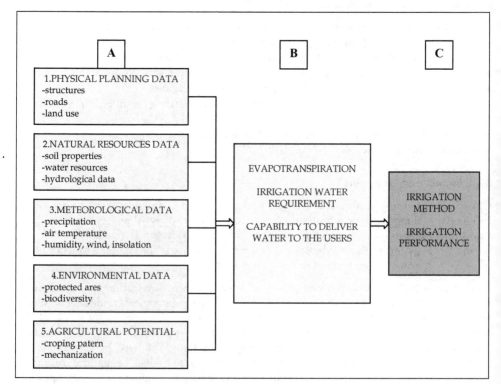

Fig. 8. Scheme of evaluation criteria and their influence on irrigation project development

According to the scheme, priorities in sustainable irrigation implementation would have already existing agricultural areas with types of soils suitable for irrigation considering their infiltration properties, areas where irrigation would not impact the overall environment. Besides, capturing of water from accessible water resources with a favourable hydrological regime must ensure sufficient amounts of water.

By meeting the required criteria, favourable conditions are accomplished for further development of a project on a lower level. Every case of neglecting of specific criteria or absence of systematic analysis leads to increase of investment value, and in long-term to overuse of natural resources and threats to the environment. On Figure 9 there is a scheme of reduction of total of agricultural land in regard to set criteria. Overall reduction may amount to over 50 %. Agricultural land found to be suitable for development of an irrigation project is subject to further economic analysis regarding application of specific irrigation methodology or choice of water sources for irrigation.

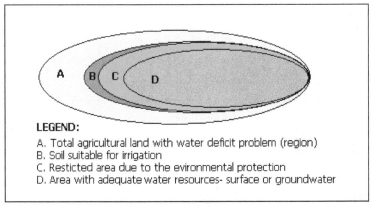

LEGEND:
A. Total agricultural land with water deficit problem (region)
B. Soil suitable for irrigation
C. Resticted area due to the evironmental protection
D. Area with adequate water resources- surface or groundwater

Fig. 9. Scheme possible reduction of total agricultural land in regard to irrigation suitability

Finally, the proposed procedure of data management contributes to the proper decision making. As an illustration, Figure 10 presents two maps of an agricultural land in Osijek County in Croatia (Master Irrigation Plan of Osijek County,2005). Previous analysis indicated a water deficit in vegetation period and frequent drought periods in the area.. The Figure 12a) shows the present state of irrigation, which is not very developed, basically only few separate fields have irrigated agriculture using water from open watercourses. After the process of evaluation of land suitability for irrigation, the second map (Fig. 10 b) shows agricultural areas with available water resources (open watercourses, ground water, reservoirs) on soil suitable for irrigation (excellent, suitable and restrictively suitable). All areas under any level of protection are considered to be unsuitable for irrigation development.

After this evaluation of land suitability for irrigation follows the procedure on a smaller scale by applying some of the multi-criteria methods. The most commonly used is linear programming, which is very suitable for this kind of problems. Optimal water management considers evaluation of available water resources in order to reach minimum expenses and satisfying needs of all other water users. Linear programming gives an optimum solution which can with minimum expenses of the system (structures, equipment, operation and

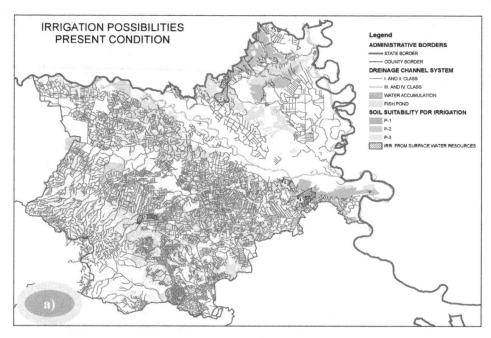

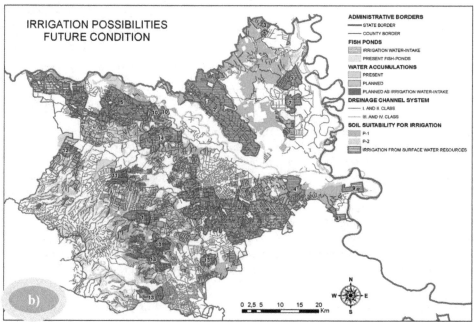

Fig. 10. Irrigation possibilities of Osijek County area, Croatia (Master Irrigation Plan of Osijek County, 2005)

maintenance) realize the maximum socio-economic benefits. Socio-economic benefits include, besides the least water price, increasing of employment possibilities, development of new economic fields, improvement of cropping pattern, etc.

10. Conclusion

To avoid subjectivity and unilaterally approach the very complex problem of irrigation implementation, methods of multi-criteria or multi-objective analysis have to be applied. This chapter tried to explain the procedure of this procedure on large-scale projects. Following of this procedure helps decision-makers to develop the project of irrigation based on sustainability and integrated water management. Considering regional physical plans, soil suitability, climatic characteristics and other geographical features, availability of water resources and their environmental vulnerability and environmental protection in general, it is possible to evaluate agricultural land suitable for irrigation. In that way, the total agricultural land will be reduced to the much smaller area, which has the good basis for irrigation implementation with reduced side effects. On a field scale, further system optimization is needed for specific cost-benefit analysis, which was not part of this elaboration.

Besides a relatively large number of potential negative side effects of irrigation described in literature and tested on the irrigated fields, without any doubts it can be said that implementation of irrigation is the necessary measure in agricultural production. The success in achieving irrigation sustainability depends on available data, reliability of the proposed procedure and reasonable data interpretation.

According to present negative trends in water and soil availability and large efforts made in environmental protection, future irrigation projects will even more depend on this kind of procedure. So we may expect development of more sophisticated and complex methods for evaluation of irrigation projects.

11. References

Ayers, R.S.,Westcot, D.W(1985). *Water Quality for Agriculture* (EDition) , FAO Irrigation and Drainage Paper 29, Rome

Bonacci, O.(2003). *Korištenje nekonvencionalnih vodnih resursa u sušnim područjima*, In: *Priručnik za hidrotehničke melioracije*, III/1, Nevenka Ožanić, 337-347, HDON, Rijeka.

Bonacci, O.(2003). *Ekohidrologija vodnih resursa i otvorenih vodotoka*, AGG Split & IGH Zagreb

Boss, M.G., Burton, M.A., Molden, D.J.(2005). *Irrigation and Drainage Performance Assessment- Practical Guidelines*, CABI Publishing, UK

Doorenbos, J.,Kassam,A.H.et.al.(1988). *Yield Response to Water* (4th E,dition) , FAO Irrigation and Drainage Paper 33, Rome

Đorđević, B.(1990). *Vodoprivredni sistemi*, Građevinski fakultet Beograd, Tehnička knjiga Beograd

Hoffman, G.J., Evans, R.G.,Jensen, M.E. et al (2007): *Design and Operation of Farm Irrigation Systems*, ASABE,

Ljubenkov, I.(2011). *Održivo gospodarenje vodnim resursima Korčule s naglaskom na njihovo korištenje za poljoprivredu*, Disertation, AGG Split, 2011.

Nixon,S.C.,Lack, T.J., HUnt,D.T.E.,Lallana,C et.al.(2000). *Sustainable use of Europe's water.* Environmetal assessment series No.7,EEA, Luxembourg

Romić, D.,Marušić,J.(2005). *Nacionalni projekt navodnjavanja i gospodarenja poljoprivrednim zemljištem i vodama u Republici Hrvatskoj (NAPNAV),* Agronomski fakultet Zagreb.

Tadić, L., Marušić, J., Tadić, Z.(2001). *Sustainable Development of Land Management in Croatia,* 19th Regional Conference of International Commision on Irrigation and Drainage,Brno-Prag

Tadić, L (2001). *Analysis of Indicators Relevant for the Sustainable Water Management on the Karašica-Vučica Catchment Area,* Građevinski fakultet Osijek

Tadić, Z. (2005). Master Irrigation Plan of Osijek County ,hidroing Osijek

Tadić, L., Tadić, Z.,Marušić, J. (2007). *Integrated use of surface and groundwater resources in irrigation,* Proceedings of 22nd European Regional Conference ITAL-ICID, Pavia, Italy

Tadić L., Bašić F.(2007). *Utjecaj hidromelioracijskog sustava navodnjavanja na okoliš (Environmental impacts of irrigation).* Proceedings of the symposium Ameliorative measures aimed at rural space development, Croatian Academy of Sciences and Arts;173-190. Zagreb

Tadić,L.,Ožanić, N., Tadić,Z., Karleuša, B.,Đuroković, Z.,(2007). *Razlike u pristupima izradi planova navodnjavanja na području kontinentalnog i priobalnog dijela Hrvatske Hrvatske vode,* Vol.15, No. 60, 201-212, Zagreb

Tadić,L.(2008). *Example of integrated use of surface and groundwater resources in irrigation in Croatia.* Journal of Water and Land Development, No. 12, 95-112

Tadić, L., Marušić, J.(2008*). Osnovni pokazatelji o načinu uređenja, korištenja i upravljanja sustavima za navodnjavanje u dijelu država Europske unije – s razvijenim navodnjavanjem.* Unpublished study.

Tadić,L.,Tadić Z.(2009). *Impacts of Irrigation on the Environment //*18th International Symposium on Water Management and Hydraulic Engineering / Editors: Cvetanka Popovska, Milorad Jovanovski. Skopje.University St Cyril and Methodius,Faculty of Civile Engineering Skopje. 2009. 659-667

Law on environmental protection (OG 82/94)

5

Rationalisation of Established Irrigation Systems: Policy and Pitfalls

Francine Rochford
La Trobe University
Australia

1. Introduction

Global concern about food security has prompted a focus on increasing productivity using increasingly scarce resources. In Australia, as in many countries, a growing population and the projected effects of climate change and shift mean that water is the focus of much of this concern. However, as in many other established economies the infrastructure designed to move water from relatively water-abundant areas to provide irrigation is aging. It requires significant investment to guard against its failure, to ensure it meets modern standards of safety, and to ensure that water is productively used and that wastage is minimised.

In accordance with the dominant global view that prioritises free trade, to which Australia subscribes, competition policy prescribes that irrigation infrastructure is provided on a 'user pays' basis – the cost of infrastructure is factored in to the cost of the water. In northern Victorian irrigation regions, where this policy has been aggressively implemented, water costs have been 'unbundled' to reflect usage, maintenance, service and infrastructure costs, so that the usage component does not form a significant part of the total water charge. In such an environment, the major infrastructure investment needed to renew irrigation infrastructure cannot be provided directly by irrigators.

The question of investment in irrigation infrastructure is common to many developed countries. In the arid states of the United States, which followed a similar 'nation-building' path in the funding of irrigation infrastructure (Lampen, 1930; Newell, 1903) irrigation infrastructure is in poor condition (US water infrastructure needs seen as urgent, 2009). Building irrigation capacity in developing countries has been identified as a priority, and major irrigation schemes are being built in China and India. However, the optimal model of investment and ownership of infrastructure is still a matter for significant debate (Abbot and Cohen, 2009). This is of particular significance when returns on commodities are subject to market distortions, so that irrigators paying for irrigation infrastructure will be competing with irrigators who are not.

The situation in Australia is of significance to any region seeking to optimise water capture and extraction for irrigation purposes. It is not clear how the current round of irrigation industry reforms will affect the industries and communities reliant on irrigation; the context, background and effects of those reforms should be closely considered.

This chapter will critically analyse irrigation industry reforms in northern Victoria, Australia. Irrigation in this region is undergoing significant organisational and infrastructure reform. With extensive assets, increasing conveyance costs and competing demands for water, managers of irrigation businesses have implemented far-reaching changes which will have flow-on effects for irrigation customers.

The social, political and legal context of these reforms is significant, so this account will commence with consideration of the national cooperative agreement that water should be managed according to the national competition agenda, the corporatisation of water authorities, the implementation of 'user-pays' principles and unbundled water charges, and the development of trade in water. Subsequent pressures as a consequence of a major drought have brought into sharp focus the environmental impact of water extractions. Market mechanisms, along with direct government acquisition of water entitlements have been directed towards reducing the irrigation 'take' from the Murray Darling system. As a consequence, in northern Victoria, modernisation, rationalisation and reconfiguration projects have been developed. These are aimed at reducing irrigation water use and contracting the coverage of irrigation infrastructure. This chapter will consider the modernisation process in northern Victoria. It will consider the implementation of a 'backbone' set of water conveyances, selected on the basis of water usage, and 'connection' back to that conveyance of channels and 'nibs'. The processes by which the 'backbone' was identified, the impact of that decision on irrigators, and the negotiation of the 'connections' program will require consideration of the formation of irrigator syndicates, the privatisation of irrigation infrastructure, and vexed questions regarding liability for failed assets, particularly on public roads and crown land. The reality of postulated water 'savings' has become a matter for political debate, and the economic and social consequences of the modernisation project have been matters of concern.

2. Irrigation policy drivers in Australia

Irrigation policy in Australia is driven by competing environmental and agricultural water needs, the cross-party acceptance of a market driven National Competition Policy (http://ncp.ncc.gov.au/), and increasing infrastructure costs as a result of aging infrastructure and increased engineering and safety requirements. These drivers impact at different levels and to different extents, and can, as will be seen, be derailed by short-term populism.

The irrigation policy environment, particularly in Victoria, has been characterised by a series of disruptive changes since the mid-1980s. These have delivered changes in governance, a decline in water availability, the introduction of water trading, unbundling of the water 'product' and change in the nature of the water 'right' itself.

2.1 Competition policy reform

Competition policy reform is expressed in states' commitment to a national competition agenda. The competition framework was endorsed by the Council of Australian Government (CoAG), made up of the Commonwealth and State Governments. In relation to water, the CoAG, an Intergovernmental Agreement on a National Water Initiative between the Commonwealth of Australia and the Governments of New South Wales,

Policy driver	Date	Primary focus
Water (Central Management Restructuring) Act (Vic).	1984	Replacement of State Rivers and Water Supply Commission with Rural Water Commission
Water Act (Vic)	1989	Introduced water trading Conversion to Bulk Entitlements
Water (Rural Water Corporation) Act (Vic).	1992	Corporatisation of Victorian water authorities
CoAG National Water Initiative	1995	Market principles increasingly applied to water
Murray Darling Basin Commission Cap	1997	Diversions from the Murray Darling Basin capped at 1993-94 levels.
Essential Services Commission Act (Vic)	2001	Regulation of rural water providers through the Essential Services Commission – introduction of a process to regulate water prices on the basis of cost recovery
Water (Resource Management) Act (Vic)	2005	Unbundling of water 'product' in Goulburn-Murray Water area - existing water rights are converted into water shares, delivery rights and water-use licences; separation of water from land; creation of water share register
Water (Governance) Act (Vic)	2006	Mirroring of corporate principles in water governance
Water Act (Cth)	2007	Federalisation of water resource management; Formulation of Basin Plan

Table 1. summary of market based policy drivers

Victoria, Queensland, South Australia, the Australian Capital Territory and the Northern Territory (CoAG, 1995) operates on the premise that national productivity will be improved by the marketisation of water resources.

The adoption of market principles is broadly consistent with view across a number of developed nations that the state provision of services is marred by state failure. State-owned enterprises were targeted for reform taking a number of forms, including structural change by unbundling activities currently provided by monopoly bodies; commercialization, by requiring that an enterprise market its services on a commercial basis to achieve at least cost recovery; contracting out of functions; corporatization to establish the body on a fully commercial basis but as a state owned company, with a delineation of the roles of the Government and the entity; and privatization of the government owned business, either wholly or partly (Department of the Treasury, 1993).

In the United Kingdom privatization of water resources occurred under the Conservative Government in 1989, representing the largest and highest level of privatization in the world. Privatization has also occurred in the United States (Water Science and Technology Board, 2002). This prevalence of the view that the adoption of market principles in the provision of government services is a good thing takes in 'a number of strands of economic thinking … claims about the nature of organizational functioning and public policy-making' (Walsh, 1995: 15) and the 'ineffective' and 'inherently wasteful' institutional framework implementing state activity and policy (Walsh, 1995: 15). Pusey notes the tendency of the dominant view to 'see the world in terms that neutralize and then reduce the norms of public policy to those of private enterprise' (Pusey, 1991: 8).

In Victoria there is a continuing reluctance to privatise water itself. The potential for a political backlash if such an attempt was made was recognised by the Victorian Parliament itself, when the *Victorian Constitution* was amended to prevent privatization of Victorian water authorities. The *Constitution (Water Authorities) Act 2003* (Vic) entrenches the responsibility of public authorities to continue to deliver water by the insertion of a new Part VII in the *Constitution Act 1975* (Vic). Section 97(1) states that if at any time on or after the commencement of section 5 of the *Constitution (Water Authorities) Act 2003* a public authority has responsibility for ensuring the delivery of a water service, that or another public authority must continue to have that responsibility. However, the section does not prevent the authority from contracting with another regarding the service, whilst retaining responsibility for it, and this has been a dominant mechanism in the provision of water services in Victoria.

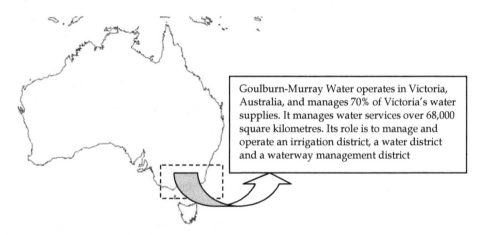

Goulburn-Murray Water operates in Victoria, Australia, and manages 70% of Victoria's water supplies. It manages water services over 68,000 square kilometres. Its role is to manage and operate an irrigation district, a water district and a waterway management district

Fig. 1. Primary case study area

The situation in Victoria is a case study for the potential effects of irrigation trade in a geographically extensive, arid, aging, largely user-pays system. The area managed by Goulburn-Murray Water (http://www.g-mwater.com.au/about/regionalmap), is comprised of gravity irrigators, pumped irrigation systems, surface water diverters, groundwater irrigators, stock and domestic customers, commercial operators (such as tourism operators), and bulk water purchasers, such as urban water corporations.

The facilitation of trade in water in Victoria is a continuation of the National Competition Policy, driven by the Productivity Commission, and now overseen by the National Water Commission (the NWC), the Australian Competition and Consumer Commission (the ACCC) and the Essential Services Commission (the ESC). According to the National Water Commission, 'water trading is a centre piece of national water reform'.

The development of a national 'grid' to enable water trade is presumed to deliver a range of benefits:

- Enhancing the capacity to 'adjust' to changing agricultural circumstances, such as low commodity prices;
- Facilitating land use changes; for instance, purchasing more water to more fully utilise land, or selling water and converting to dryland farming;
- Enabling irrigators to 'hedge' against periods of poor rainfall by selling irrigation entitlements; for instance, in a period of prolonged drought some irrigators are able to convert to non irrigation operations, or to
- Liberating the embedded capital in a farming enterprise by enabling irrigators to sell or secure against their entitlement (separately from land);
- Enabling the transfer of water to urban use, thus adding economic value to regional communities;
- Improving the efficiency of water use by enabling water delivery to reflect market costs and encourage transfer to more efficient uses.

Water trading in major irrigation regions in Victoria can occur on a permanent or a temporary basis, and sophisticated methods have been devised to facilitate its occurrence. Water brokerage firms are common, and large water suppliers have user-interface systems which allow online trading of water. For instance, Goulburn-Murray Water developed the 'Watermove' web trading interface at www.watermove.com.au. It allows trade in water allocation, water shares, groundwater, and unregulated surface water. There is no doubt that water trading has, since its inception, provided significant flexibility for irrigators. During the recent decade-long drought water trading was particularly beneficial, and analysts of the effect of water trading have used these figures to illustrate its positive effect (National Water Commission, 2010a). A clear picture of the effects of water trade in an individual irrigation area is more problematic; consolidated records provide detailed information about the amount of water traded, whether it is permanent or temporary, high security or low security water, and the regions from or into which it was traded (National Water Commission, 2010b).

An indicative comparison of water traded in the Goulburn system in the height of the recent drought (2007 – 2010 irrigation seasons) demonstrates relatively flat net volumes traded (into the irrigation district) but high total volumes traded, demonstrating that water was moving between irrigators. It is possible to interpret this as a rational response to water shortage by those able to obtain a higher price per megalitre of water by selling it than they could by utilising it - particularly since overall water allocations may have been too low to continue normal farming operations. The fluctuations in temporary price are more indicative of the yield per megalitre of water, since purchase of permanent water in a given year would not necessarily deliver a full water allocation in that year.

Positive messages about the effect of water trade as a comparison to the alternative – water attached to land and unable to be traded - display some blindness to the historical position. Because of State government policies prior to the marketisation of water, irrigators were required to pay for water regardless of whether they received it or not. This was deliberate measure to ensure the ongoing viability of the irrigation infrastructure, and a lesson learned by government by the failure of private irrigation ventures during years of drought. Prior to 1986, during years of low allocation, irrigators could receive no water, but still be required to pay for that water. Conversely, if they did not need the water, but received an allocation, they were obliged to pay for it regardless. Thus, there was no incentive to conserve water, severe impediments to changing land use, and ongoing costs in years of low income. The capacity to trade an allocation delivered immediate benefits. Irrigators whose land use was constrained by lack of water could improve the productivity of that land by purchasing additional water, and farmers who wished to transition out of farming, or transition out of irrigation farming, had additional mechanisms with which to do so.

Trading in the Goulburn 1A system 2007-11

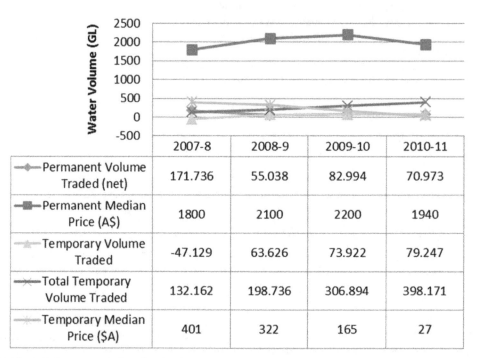

	2007-8	2008-9	2009-10	2010-11
—◆—Permanent Volume Traded (net)	171.736	55.038	82.994	70.973
—■—Permanent Median Price (A$)	1800	2100	2200	1940
—▲—Temporary Volume Traded	-47.129	63.626	73.922	79.247
—✕—Total Temporary Volume Traded	132.162	198.736	306.894	398.171
—┼—Temporary Median Price ($A)	401	322	165	27

Table 2. Volumes traded in the Goulburn 1A Irrigation Zone.

However, marketisation of water was accompanied by a number of other mechanisms, including the unbundling of the water product into a number of products representing separate charges for delivery, infrastructure, and water components. The actual volume of water is not a major proportion of the bill. Thus, under the provisions of the *Water (Resource Management) Act* 2005 (Vic), which amends the *Water Act* 1989 (Vic), existing water rights were converted into water shares, delivery rights and water-use licences. The irrigator is able to trade the actual water share, but the infrastructure access fee would still be payable, unless the irrigator surrenders it. In order to surrender the access fee the irrigator has to pay a termination (exit) fee, which can be prohibitively expensive (ACCC, 2009b).

The purpose of the infrastructure access charges and the termination fees is to ensure the viability of the infrastructure in the event of significant numbers of water users exiting the water district. The obvious corollary to this is that, contrary to the principle of facilitating flexibility in land and water use, holders of large delivery shares are locked into irrigation enterprises. There is the theoretical potential for people to trade the delivery share, however there is no market for that component. During the unbundling process irrigators with an existing right were given one delivery share for each hundred megalitre of water entitlement. However, the delivery share was devalued because irrigators requiring temporary water can acquire 270 megalitres on a parcel of land for each delivery share. There is, therefore, no market for the sale of delivery shares.

2.2 Environmental pressures

The environment is traditionally a matter within state Constitutional competence. State-based environmental legislation has a significant impact on the delivery of water in rural areas. Longstanding environmental measures at state level include those pursuant to the *Catchment and Land Protection Act* 1994 (Vic), the *Environment Protection Act* 1970 (Vic), the *Flora and Fauna Guarantee Act* 1988 (Vic), the *Heritage Rivers Act* 1992 (Vic) and the *Planning and Environment Act* 1987 (Vic). Environmental requirements also apply under Part 3 of the *Water Act* 1989 (Vic). Coverage by federal legislation has increased as a consequence of High Court interpretations of the external affairs power enabled by s.51(xxix) of the constitution, but significant co-operative measures had been taken to implement desirable environmental measures. In particular, the Murray Darling Basin Commission Cap was implemented to restrict diversions to 'the volume of water that would have been diverted under 1993/94 levels of Development. In unregulated rivers this Cap may be expressed as an end-of-valley flow regime' (MDBMC, 1996). The primary objectives of implementation were:

1. to maintain and, where appropriate, improve existing flow regimes in the waterways of the Murray-Darling Basin to protect and enhance the riverine environment; and
2. to achieve sustainable consumptive use by developing and managing Basin water resources to meet ecological, commercial and social needs.

As a consequence of the environmental stresses occasioned by the recent drought there has been a wholesale attempt to federalise basin-wide management of the water resource. The *Water Act* 2007 (Cth), partially based on a patchwork of constitutional powers and partially a result of negotiations between the Commonwealth and each Basin State, was realised only when a drought of over a decade duration began to threaten urban water security. However, it had as its fundamental premise the desire to manage the Basin on a global basis, and in particular to limit extractions of water, in order to 'provide for the integrated management of the Basin water resources in a way that promotes the objects of [the *Water Act* 2007 (Cth)], in particular by providing for:

a. Giving effect to relevant international agreements (to the extent to which those agreements are relevant to the use and management of the Basin water resources); and
b. The establishment and enforcement of environmentally sustainable limits on the quantities of surface water and ground water that may be taken from the Basin water resources (including by interception activities); and
c. Basin-wide environmental objectives for water-dependent ecosystems of the Murray-Darling Basin and water quality and salinity objectives; and
d. Water to reach its most productive use through the development of an efficient water trading regime across the Murray-Darling Basin; and

e.   Requirements that a water resource plan for a water resrouce plan area must meet if it is
     to be accredited or adopted under Division 2; and

f.   Improved water security for all uses [sic] of Basin water resources (*Water Act* 2007 (Cth)
     s.20).

The Basin Plan has not yet been released; at the time of writing it had been delayed again
until October 2011 (Slattery, 2011). Significant controversy has arisen over the appropriate
balance to be struck between environmental, social and economic values in devising the
Plan (ABC News 2010; Stubbs, Storer, Lux and Storer 2010). Whether the environment was
to have priority in the final Basin Plan was a matter of competing legal views. There was a
real question as to whether the Act required the Authority to privilege the environment over
other concerns (Kildea and Williams, 2011). There are significant concerns as to whether the
Authority is the appropriate body to balance social and economic factors with
environmental concerns. The forwarding of social objectives is more properly left to political
consideration. The *Guide to the Proposed Basin Plan* (MDBA, 2010a) prioritised the
environment and required significant cuts to irrigation entitlements, but the negative
response to the proposed plan (Cooper 2010; Lloyd, 2010a), however, and the return of rain
(Lloyd 2010b) have delayed the progress of reforms.

2.3 Regional policy

Regional policy frequently demands political responses, and the vulnerability of policy-
making which affects rural communities was demonstrated by the political fallout from the
Guide to the proposed Murray-Darling Basin Plan (MDBA, 2010a). The priority for regional
policy in Australia has, however, for many years, been the facilitation of 'sustainable' or
'resilient' communities, and the 'adjustment', with government assistance, of those that
appear to be unsustainable. The government or quasi-governmental agency enables the
individual or community to become a self-sufficient agent. Marketisation of water
infrastructure and water resources is consistent with this view, since it conceptualises the
individual as capable of utilising transactional mechanisms, such as contract, to achieve
optimal personal outcomes. Full cost recovery on government supplied infrastructure, such
as dams and channels, is necessary to ensure that the community is 'sustainable'. Trade in
water ensures that water can move from an 'unsustainable' community to a sustainable one.

Analyses of the operation of market mechanisms for water transfer have been characterised
as supporting this view; the National Water Commission, in a study of the effects of water
trading in the southern Murray Darling Basin, concluded that 'water markets and trading
are making a major contribution to the achievement of the NWI objective of optimising the
economic, social and environmental value of water. The overwhelming conclusion of the
study is that water trading has significantly benefited individuals and communities across
the sMDB' (NWC, 2010, v).

The interaction between regional policy and the various water policies, however, is complex,
particularly where the contraction of essential infrastructure is concerned. The basis upon
which infrastructure – particularly water infrastructure – is reduced has far-reaching
consequences for regions, since it affects rural rate bases, school and hospital viability, and a
range of other service that depend on population density. The contraction of water
infrastructure in the northern Victorian irrigation regions is driven by the decision that the

infrastructure is unsustainably expensive on a user-pays basis. As a generalisation this is problematic; there are elements of cross-subsidisation across irrigation districts in the larger water suppliers. Like all organisations the depreciation of infrastructure and the allocation of maintenance costs make a significant difference to whether the area is operating at a loss. Overall, the entirety of Goulburn-Murray Water is required to operate on a full-cost recovery basis (Standing Committee on Finance and Public Administration, 2007 - 2008). However, one of the consequences of modernisation of infrastructure will be an increase in the cost of that infrastructure for users on an ongoing basis.

3. Political, and environmental stressors

3.1 Water scarcity and centralisation of water policy

Thus, at the commencement of the new millennium Australian water policy was broadly consistent. However, the manner by which states implemented that policy, and the degree of compliance with key objectives, varied significantly. The more populous states of New South Wales and Victoria, with developed irrigation industries and a long history of appropriation, do not have the same interests as Queensland, with a shorter history of development and greater incentive to continue to allow diversions, or South Australia, with a capital city entirely dependent on extraction and a developing horticultural industry. Although compliance with national water policy is assessed, and Commonwealth tranche payments are dependent on that compliance, states' ability or willingness to set up the appropriate mechanisms has not always been evident. The Commonwealth suspended competition payments to SA, Victoria and NSW for not meeting their commitment to enable interstate trade in water by the agreed date of July 2006, and subsequently gained in-principle agreement to enable trade.

The decade of drought in the 1990s, however, allowed the federal government to assert significant political pressure on the states to centralise water resource management in the basin more thoroughly. Increasing pressure on urban supplies necessitated massive investments in infrastructure to ensure continuing supply of water to major population centres. The perennial state shortage of infrastructure funds for the construction of pipelines and desalination plants to augment city supplies was answered by federal government leverage of funding to obtain agreement to the referral of powers necessary to pass the federal *Water Act* 2007 (Cth). The explanatory memorandum to the Act stated that it gives effect to a number of key elements of the Commonwealth Government's $10.05 billion *National Plan for Water Security*. The Act is intended to enable water resources in the Murray-Darling Basin to be managed in the national interest, optimising environmental, economic and social outcomes.

The Act was contentious, and subject to constitutional challenge, particularly by the State of Victoria, which had initially refused to refer its powers to the Commonwealth. The Act in its final form commenced operation on 3 March 2008; however, as stated above, the Basin Plan required under the Act has not yet been finalised.

3.2 State policy and the rush for results

The availability of federal funding was a significant incentive, particularly for Victoria. However, the Victorian government would have suffered political backlash if it had simply built a pipeline from rural water supplies and purchased water to augment urban supplies. Instead, it took the opportunity afforded by a proposal by rural interests to 'save' water by improving rural infrastructure, paying for it with a combination of urban, state and federal government money, and splitting the water 'saved' between the environment, urban water consumers and rural water consumers. Problematically, however, these decisions were made under the pressure of a drought, and urban water security, particularly in the capital city of Melbourne, was threatened. Thus there was significant pressure to find water 'savings' quickly, and expedite infrastructure development.

4. The 'modernisation' of irrigation infrastructure

4.1 Water savings and trade-offs

The major support for funding of infrastructure improvements is through identification of water 'savings' that can be deployed as environmental water or 'sold' to urban use. The trade metaphor enables consistency with the market-based premise within which the industry operates. Water 'savings' are generated through the replacement of meters, installation of regulators to enable closer regulation of the channel system to prevent outfall and reduce seepage (by running the channel lower) and in some cases through the lining or piping of channels. Leaving aside the highly questionable assumption of defective Dethridge wheels inevitably measuring in favour of farmers, for which significant savings have been claimed, the most significant water savings are generated through the privatisation or retirement of irrigation infrastructure.

The technologies of performance to monitor savings have, unfortunately, been preceded by the projects they are meant to be monitoring. Thus, audit of outcomes of the rationalisation processes has been performed without baseline data and the protocols for quantification of water savings had to be developed after the actions upon which the water savings were dependent had been commenced (DSE, 2010).

An early analysis of the use of water 'savings' mechanisms over other techniques was carried out by the Productivity Commission, which noted that

> One of the purported benefits of water saving investment over market purchase is that it avoids reductions in rural water use by creating 'new' water. However, water 'savings' associated with indirect purchases can be illusory. That is, measures to reduce system losses actually divert water from other beneficial uses, elsewhere in the system, that rely on return flows (PC 2006b). For example, total channel control is a water delivery technology that uses automated control gates to reduce irrigation district outfalls and improve service quality. However, district outfalls often supply downstream water users. Transferring entitlements out of the system based on illusory water savings can therefore 'double up' losses in return flows (Productivity Commission, 2008: 78)

The Productivity Commission reviewed the operation of the NVIRP alongside other water purchase mechanisms to product environmental flows – largely tender mechanisms - in 2009. However, consistent with the Commission mandate it was predicated on the use of market mechanisms to achieve the environmental outcomes and was primarily concerned with the interaction of proposed mechanisms (NVIRP, 2009).

4.2 Contraction of irrigation infrastructure

The Northern Victorian Irrigation Renewal Project (NVIRP) is a state-owned entity; the Chief Executive Officer reports through the NVIRP Board to the Minister for Water and the Treasurer. When the modernisation process is complete the assets constructed will be transferred to Goulburn-Murray Water. The aim of the 'modernisation' project now being implemented by NVIRP is to deliver 'a more efficient and affordable irrigation delivery network that is able to deliver an improved level of water delivery service and increase on-farm productivity' (DSE, 2004). Water 'savings' to be delivered from this program have been estimated as up to 425 GL annually. Of that amount, 75 GL were intended to be diverted to Melbourne, 175 allocated to the environment and 175 to irrigators in the system (King and Tonkin, 2009). The program was to have been partially funded by Melbourne Water premised on the diversion of water to Melbourne, but this was subsequently changed (Victoria Auditor General, 2009, vii). The 'Core Principles' of NVIRP, espoused by the Food Bowl Modernisation Project Steering Committee report and endorsed by the Victorian Government on 30 November 2007, are to:

- Focus on economic development
- Strive for efficiency in both water supply and farm watering systems
- Provide different levels of service to meet the needs of different customers and customer types
- Strive for an on-demand water delivery service
- Develop system components that ensure cost and service competitiveness in water supply
- Develop policies to support and guide decisions
- Stage project delivery to match funding availability (NVIRP, (nd b)).

The majority of these principles demonstrate the

'post-welfarist regime of the social' in which 'performance government' displaces the collectivist ethos of welfarism. Here, various state and non-state agencies become facilitators both in optimizing individual capacities to act in an entrepreneurial *and* socially responsible way, and in the diagnosis of potential risks that threaten to disrupt the achievement of personal liberty (Higgins and Lockie, 2002, 421).

The NVIRP not only reconstructs government infrastructure (on a user-pays basis) it will facilitate on-farm irrigation works, funding them with water off-sets, to enforce efficiency gains. Irrigators will inherit a high-functionality, high-cost irrigation network, and a fully marketised water trading system will enable the transfer of water from those unable to afford the higher water costs to higher value uses – such as urban use.

The irrigation area involved in the project is around 800,000 ha and 14,000 farms (Spencer, 2010: 17), and by any measure the injection of funds into the project is significant. Around $2 billion is projected to be utilised in the project in replacement of meters, and regulators lining channels and implementation of 'total channel control' systems. 'Total Channel Control' is a Rubicon Systems product aiming for 'end-to-end irrigation canal automation technology ...[and] transforming the inefficient manually operated open canal networks into fully automated, integrated and remotely controlled systems that are achieving demonstrated new benchmark delivery efficiencies of up to 90%' (Spencer, 2010, 19). The

90% efficiency claim is substantiated by a reference to the Coleambally Irrigation district during the 2006-07 year (a drought year during which irrigators had a 10% irrigation allocation in that system) (DEWR, 2007) and refers to the claims that irrigators were allocated an extra 18% water allocation because of 'savings' from the new system; but it is not clear whether this is carryover from the previous year. It is not clear whether the 'benchmark' of 90% is an average or a measure on one channel in the system. It is being compared with the 73% *average* across irrigation systems. Losses will vary according to a range of factors including soil type, gradient, supply level and infrastructure age. Further, Coleambally was a greenfields site; the infrastructure was installed when the district was developed. Thus, it was new infrastructure, and the costs of retrofitting century-old infrastructure was not an issue.

Commentators have lauded the effects of the modernisation; the Business Development Manager of Rubicon Systems (which supplies the 'Flumegates' and 'Total Channel Control' mechanisms for the upgraded system), has reported that

> A higher level of technological investment in the modernization of large unlined, gravity fed irrigation systems in the south-eastern state of Victoria has resulted in increases in efficiency from about 70% up to about 90% - a remarkable outcome. In Victoria, the water saved is being reallocated equally between urban and industrial users, the environment and to existing farmers to improve their security of supply (Spencer, 2010: 15).

Other commentators have been less laudatory; the cost of the infrastructure program in delivering environmental and urban water far exceeds the cost of purchasing the water on the market. Some irrigators are concerned not only about the projected contraction of the irrigation system, but also about the potentially unsustainable cost of the new technology, which will be likely to have a shorter lifespan than previous low-technology solutions such as the Dethridge Wheel, and will require them to bear higher ongoing infrastructure charges.

4.2.1 The backbone

The most significant factor in generating the savings required by the Northern Victorian Irrigation Renewal Program will be the contraction of irrigation infrastructure to the 'backbone'. This is the network of channels closest to the main carrier, based on the delivery share on that particular channel. Thus, modernisation works are being carried out to service those farms on the 'backbone'.

The contraction of irrigation infrastructure to the backbone was prefaced in the Victorian White Paper:

> Rationalisation of services is primarily an issue for north-central Victoria. Goulburn-Murray Water and its water service committees realize that some parts of existing distribution systems need to be closed down. They were constructed in an era of bold development and in some places are just too spread out, as well as being on land that has turned out to be unsuited to irrigation (DSE, 2004: 82)

Rationalisation was originally a separate government program, but it appears that it has now been rolled over into the Northern Victorian Irrigation Renewal Program, resulting in difficulties ascertaining whether the objectives of either program had been met.

Rationalisation of irrigation infrastructure has not been confined to 'land that has turned out to be unsuited to irrigation.' The first irrigation district to be closed was the Campaspe Irrigation District, on excellent land and close to a natural carrier, and, ironically, flooded in early 2011 and in the following season water entitled to 100% allocation. The backbone becomes the de facto mechanism for limiting public funding of infrastructure. Those on the backbone undergo a series of consultation mechanisms to determine their current and future business needs – the farm irrigation assessment process – after which a decision is made by NVIRP as to their infrastructure requirements to meet those needs. The infrastructure will be installed and monetary compensation will be paid on the basis of assets removed. Additional programs utilising federal money and handled through the Department of Primary Industries finance on-farm efficiency works such as the installation of pipes and rises to replace flood irrigation, the piping of on-farm channels, and the laser levelling of land in return for the irrigator surrendering water. The agencies are therefore facilitating the projected business infrastructure requirements – they have taken on an enabling role, brokering deals to increase the efficiency of the irrigation operations of the farm.

4.2.2 The connection programs

The primary 'technology of agency' is the connections program, pursuant to which individuals or groups who are not on the backbone must negotiate either alone or with neighbours to connect to the backbone. NVIRP notes that

> NVIRP's Connections Program involves connecting irrigators to a modernised main system of irrigation channels or 'backbone'. The program aims to consolidate supply point connections and ensure as many customers as possible are connected directly to the backbone to access improved water delivery services.

> Properties are connected to the Goulburn-Murray Water channel supply system via supply point connections. Through the Connections Program, irrigators are being encouraged to upgrade their supply point connections or move supply points from secondary or spur channels to the backbone via a new connection, adopting the solution that best suits their farming operations (NVIRP, nd c).

This may mean that monetary incentives for connection are available based on water savings. Further incentives are available for on-farm efficiency works from programs like the Farm Water Program (Goulburn Broken Catchment Management Authority, 2010). NVIRP processes are mediated first by negotiation between one or a number of landowners, then by contract: a standard Rationalisation Agreement (NVIRP, nd a) forever discharges Goulburn-Murray Water and NVIRP 'from any and all claims and rights for any cost, loss, liability, damage, compensation or expense arising out of or in connection with the Rationalisation or the matters contemplated by this Agreement' (NVIRP, nd a: para 8(b)). Upon signature, the landowner accepts payment of compensation 'in full and final satisfaction of all claims…in connection with the matters contemplated by the Agreement' (NVIRP, nd a: para 7(b)).

The overall difficulty with the connections program at this stage, however, is the ongoing uncertainty for irrigators who have found themselves *off* the backbone, even though their farming enterprises are otherwise sustainable and profitable. Although the program rollout has occurred over a number of years, the lack of detail on the manner in which 'connections' to the backbone will occur has been problematic, partly because it introduces the issue of the privatisation of infrastructure, and the risks and losses associated with infrastructure.

4.2.3 Privatisation of risks and losses

The perennial debate about the efficacy of 'market' mechanisms for the delivery of public services attracts the usual criticisms of wasteful government service provision (Brody, 2005: 3). Conversely, critics of market mechanisms as primary devices for delivery of social obligations instance failures in the market due to monopolization of private providers, failure to provide adequate incentives for delivery of social obligations and rising prices after privatisation of government services. Since instances of problematic introduction of market mechanisms can be dismissed where benchmarks for successful private delivery were insufficiently defined, and since reintroduction of full government provision of many services is not on the agenda, the debate must be more strategically defined.

The connections program brings public infrastructure to a single supply point. Although details have not been finally determined, the dominant model has been that water will be metered at that point. From that point, infrastructure requirements are privatised. Ongoing maintenance of that infrastructure is the obligation of the landowner or group of landowners. Additional infrastructure, such as road culverts and bridges, were also anticipated as included in private obligations, but local councils have expressed disquiet with that arrangement, and refused to accept applications for planning approval, and it has now been indicated that water authorities will be required to maintain responsibility for these assets.

If more than one landowner requires water from that single supply point, the administration of water from that supply point is also a matter for private negotiation. These arrangements are considered to be primarily of a commercial nature; decisions will be based on the current and projected irrigation business needs. This also privatizes losses on the infrastructure below the supply point, and enables NVIRP to claim these losses as part of the program savings. The expectation appears to be that contractual mechanisms will mediate relationships between affected irrigators.

Those not on the backbone continue to meet the other infrastructure demands of the system. Thus, landholders pay delivery share on the basis that they remain connected. Further, irrigators will be constrained from 'exiting' the system without the requirement to pay an exit fee. The ACCC oversees the obligation to pay an exit fee (ACCC, 2009a). The maximum termination fee allowed by the ACCC is 10 times the total infrastructure access fee (although a common requirement to pay the current year's infrastructure access fee makes it, in reality, 11 times that fee), which equates to the amount payable per delivery share. For many irrigators of average size, this amount will be in the hundreds of thousands, and it will vary between irrigation areas because the cost of maintenance of infrastructure will vary between areas. Ironically, those areas which have been most extensively modernised and thus have the most expensive infrastructure may also have the highest ongoing costs, and thus be the least sustainable. The issue of cross-subsidisation between modernised and unmodernised areas should now be closely monitored.

The ACCC notes that this arrangement may be varied by agreement, and NVIRP has indicated publicly that irrigators will be permitted to negotiate exit fees. However, where connections programs have not commenced this has not been an option.

Currently, as the details of the connections program have not yet been finalised, many irrigators are unsure whether connection back will be a viable alternative, as it will force irrigators to bear high infrastructure and maintenance costs and to reach negotiated positions with multiple irrigators without the statutory scaffolding available to authorities.

5. Conclusions

The political arguments through which the contraction of infrastructure has been made more palatable have been arguments for 'modernisation', efficiency, and the return of water to the environment. These have been most compelling during periods of water scarcity, and have been utilised to ensure that the grave political consequences of the failure of urban infrastructure have not eventuated. The processes through which modernisation have occurred have had the effect of privatising significant portions of infrastructure and transferring risk from the state to an individual or group of individuals.

There are significant risks in forwarding this strategy for regions. The increasing costs of maintaining a water supply, along with the potential to trade water to alleviate those costs, result in more and more water leaving irrigation districts. Since land with water produces more and can support greater numbers of people, the consequence of water leaving land tends to be an overall reduction in the economic wealth of the community and a reduction in the number of people in that community.

However,

> The distribution effects of water trade depend on whether the people who sell the water stay in the region and whether they invest outside the region. The effects will also depend on whether those purchasing temporary allocations are doing so to offset their sale of entitlements, or whether those irrigators selling entitlements are different to those who are purchasing allocations.

> The overall trend across GMW's main irrigation districts was a decline in the number of people employed in Agriculture, Forestry and Fishing … by 5% between 1996 - 2001. (DSE, 2008).

The decline in rural populations is frequently considered to be an inevitable consequence of the economic conditions in first world countries, and concerns about food security (for instance, O'Grady 2011; Schmidhuber and Tubiello, 2007; Brown and Funk 2008), are alleviated by the argument that modern farming conditions, being more efficient, require fewer participants (c/f Altieri and Rosset, 2002). However, the consequence of the contraction of irrigation is an exacerbation of loss of population by a loss of infrastructure. This is a removal of both the industry and the capacity for the industry to continue.

6. Acknowledgements

Figures in Table 2 are compiled from the Department of Sustainability and Environment, Victorian Water Register website at http://waterregister.vic.gov.au/Public/Reports/WaterAllocation.aspx

7. References

ABC News (2010) 'Report says irrigation cuts to slash jobs', 6 August [Online] Available: <www.abc.net.au/news/stories/2010/08/06/2975836/htm> [2010, August 9]

Abbot, M., Cohen, B., (2009) 'Productivity and efficiency in the water industry'. 17 *Utilities Policy* 233-244

Altieri, M.A. and Rosset, P., (2002) 'Why biotechnology will not ensure food security, protect the environment, or reduce poverty in the developing world' in Sherlock, R. and Morrey D.J. *Ethical issues in biotechnology* 175 – 182

Australian Competition and Consumer Commission (ACCC), (2009a) *Permanently selling your water and terminating your delivery right – a guide for irrigators about the water market rules and rules on termination fees.*

Australian Competition and Consumer Commission (ACCC), (2009b) *Water charge (termination fees) rules: technical guide for operators* June.

Brody, G., (2005) 'Executive Summary', in Consumer Law Centre Victoria and Environment Victoria, *Water: Access, Affordability and Sustainability: Issues Paper* May

Brown, M.E. and Funk C.C., (2008) 'Food security under climate change' 319 *Science* 580 – 581

CoAG, (1995). Intergovernmental Agreement on a National Water Initiative between the Commonwealth of Australia and the Governments of New South Wales, Victoria, Queensland, South Australia, the Australian Capital Territory and the Northern Territory, at

http://www.coag.gov.au/meetings/250604/index.htm#water_initiative

Cooper, H., (2010) 'Murray–Darling plan gets fiery reception,' *Lateline Australian Broadcasting Corporation,* 14 October

www.abc.net.au/lateline/content/2010/s3038842.htm (accessed November 17, 2010).

Department of Environment and Water Resources (DEWR), (2007), *Innovation in Irrigation – case studies from across Australia* http://www.nht.gov.au/publications/case-studies/irrigation2007/pubs/irrigation2007.pdf

Department of Sustainability and Environment (DSE), (2004), *Our Water Our Future: Securing our water future together,* State of Victoria, White Paper

Department of Sustainability and Environment (DSE), (2008), Water Trading in Northern Victoria 1991/92 - 2005/06

Department of Sustainability and Environment (DSE), (2010), 'Water savings protocol: technical manual for the quantification of water savings' Version 3 August

Goulburn Broken Catchment Management Authority, (2010), 'Farm Water Program' http://www.gbcma.vic.gov.au/default.asp?ID=farm_water

Higgins, V., Lockie, S., (2002), 'Re-discovering the social: neo-liberalism and hybrid practices of governing in rural natural resource management' 18 *Journal of Rural Studies* 419-428

King, A.J., and Tonkin, Z., (2009), 'Northern Victoria Irrigation Renewal Project: Operational impact assessment on aquatic fauna', Unpublished client report, Arthur Rylah Institute for Environmental Research. Department of Sustainability and Environment, Heidelberg available at http://nvirp.spyro.ddsn.net/downloads/Planning/PER/App_11._NVIRP_Assessment_of_aquatic_species.pdf.

Lampen, D., (1930) *Economic and Social Aspects of Federal Reclamation* Baltimore: The Johns Hopkins Press

Matthews, K., (2008), Chair and Chief Executive Officer, National Water Commission, *Australian Water Markets Report 2007- 2008*, December, National Water Commission, Canberra,

Kildea, P., and Williams, G., (2011), 'Professors Paul Kildea and George Williams, Gilbert and Tobin Centre of Public Law, UNSW' *Submissions received by the Committee*, 12 January 2011. http://www.aph.gov.au/senate/committee/rat_ctte/mdb/submissions.htm (accessed April 26, 2011).

Lloyd, G., (2010a), 'Great dividing rage over water' *The Australian*, 23 October. <www.theaustralian.com.au/news/opinion/great-dividing-rage-over-water/story-e6frg6zo-1225942074914> (accessed November 17, 2010).

Lloyd, G., (2010b), 'The drought breaks' *The Australian*, 13 November. www.theaustralian.com.au/national-affairs/the-drought-breaks/story-fn59niix-1225952583089 (accessed November 13, 2010).

Murray–Darling Basin Authority (MDBA), (2010a), *Guide to the Proposed Basin Plan* [Online] Available: http://thebasinplan.mdba.gov.au/guide/

Murray-Darling Basin Ministerial Council, (1996), *Setting the Cap – Report of the Independent Audit Group* November

National Water Commission (NWC), (2010a), *The impacts of water trading in the southern Murray–Darling Basin: an economic, social and environmental assessment*, NWC, Canberra

National Water Commission (NWC), (2010b), *Australian Water Markets Report 2009-2010* available at http://www.nwc.gov.au/www/html/2971-water-markets-report---december-2010.asp?intSiteID=1

Newell, F.H., (1903) 'The Reclamation of the West' *Annual Report of the Smithsonian Institution*

NVIRP, (2009), Submission to the Productivity Commission Review of Market Mechanisms for Recovering Water in the Murray-Darling Basin 8 September 2009 available at http://www.pc.gov.au/__data/assets/pdf_file/0009/91494/sub038.pdf

NVIRP (nd a), 'Rationalisation Agreement', available at http://www.nvirp.com.au/downloads/Connections/Connections_Program/Sample_Legal_Agreement.pdf

NVIRP (nd b) 'Vision and principles' http://www.nvirp.com.au/about/vision_principles.aspx

NVIRP (nd c) http://www.nvirp.com.au/connections/connections_program.aspx

Office of State Owned Enterprises, Department of the Treasury, (1993), *Reforming Victoria's Water Industry: A Competitive Future*

O'Grady, S., (2011), 'The coming hunger: Record food prices put world 'in danger', says UN' *The Independent* 6 January [online] http://www.independent.co.uk/news/world/politics/the-coming-hunger-record-food-prices-put-world-in-danger-says-un-2177220.html [accessed January 6, 2011]

Michael Pusey, (1991), *Economic Rationalism in Canberra: A Nation-Building State Changes its Mind* Sydney: University of New South Wales

Murray-Darling Basin Authority, (MDBA), (2010), *Guide to the proposed Basin Plan*. Canberra: Commonwealth of Australia

National Water Commission, (2010), *The impacts of water trading in the southern Murray-Darling Basin: an economic, social and environmental assessment*, NWC, Canberra

Productivity Commission, (2008). *Towards Urban Water Reform: A Discussion Paper*. Productivity Commission Research Paper, Melbourne, March

Schmidhuber J., Tubiello, F.N., (2007) 'Global food security under climate change' vol. 104 no. 50 PNAS 19703–19708

Slattery, C., (2011), 'Draft Murray Darling plan delayed' *ABC News* [online] August 3

Spencer, G., (2010) 'Making large gravity irrigation systems perform' in CA Brebbia, AM Marinov and H Bjornlund (eds) *Sustainable Irrigation Management, Technologies and Policies III* 134 WIT Transactions on Ecology and the Environment 15

Standing Committee on Finance and Public Administration, (2008), *Report on Goulburn-Murray Water's Performance 2007-8*

Stubbs J., Storer J., Lux C. and Storer T., (2010) *Exploring the Relationship Between Community Resilience and Irrigated Agriculture in the MDB: Social and Economic Impacts of Reduced Irrigation Water*, report (Stubbs Report), 4 July [Online] Available: <www.cotton.crc.org.au/content/General/Research/Projects/3_03_05.aspx> [2010, August 9]

Turnbull, M. (2006), 'Malcolm Turnbull Speaks to the National Press Club in Adelaide' (Speech to the National Press Club, Adelaide, 22 November 2006) (<http://parlinfo.aph.gov.au/parlInfo/download/media/pressrel/4VLL6/uploa d_binary/4vll63.pdf;fileType%3Dapplication%2Fpdf> at 10 August 2011 (copy on file with author).

'US water infrastructure needs seen as urgent' (2009) *Reuters* [online] May 8, available at http://www.reuters.com/article/idUSTRE54731G20090508?loomia_ow=t0:s0:a49:g 43:r1:c1.000000:b30495740:z0, 17 June 2010.

Victorian Auditor General, (2009), *Water Entities – Results of the 2008-09 Audit* November 2009

Walsley, J., Argent, N., Rolley, F., Tonts, M., (2006), 'Inland Australia' (Paper presented at *Australians on the Move: Internal Migration in Australia*, the 2006 Annual Symposium of the Academy of Social Sciences in Australia, Australian National University, Canberra, 20–22 November 2006). Abstract avail <http://www.assa.edu.au/Conferences/details/symp2006.htm> at 27 November 2006 (copy on file with author).

Walsh, K., (1995), *Public Services and Market Mechanisms: Competition, Contracting and the New Public Management* Palgrave Macmillan

Water Science and Technology Board, (2002) *Privatization of Water Services in the United States: An Assessment of Issues and Experience*

Part 2

Strategies for Irrigation Water Supply and Conservation

6

Optimal Design or Rehabilitation of an Irrigation Project's Pipe Network

Milan Cisty

Slovak University of Technology Bratislava
Slovak Republic

1. Introduction

This chapter deals with the optimal design of the hydraulic part of an irrigation project. It is mainly focused on the design or management of the sprinkler irrigation method.

A typical sprinkler irrigation system usually consists of the following components:

- Source of the water (reservoir, river, well, waste water)
- Conduit to irrigation area (for instance, a canal or pipe system)
- Pump unit
- Pipe network (mainline and submainlines)
- Sprinklers or various types of irrigators (center pivot sprinkler system, linear move system, traveling big gun system, portable hand-move lateral pipe system, solid-set irrigation system, etc.)

This chapter deals with the optimal design of the most expensive part of a pressurised irrigation system, i.e., its pipe network. The problem of calculating the optimal pipe size diameters of an irrigation network has attracted the attention of many researchers and designers. During the process of designing an irrigation pipe network, the hydraulic engineer will face the problem of determining the diameters of the pipes forming the distribution network. For economic reasons, the pipe diameter should be as small as possible; on the other hand, the diameter must be large enough to ensure service pressure at the feed points (Labye et al., 1988; Lammadalena and Sagardoy, 2000).

Many optimization models based on linear programming (LP), non-linear programming (NLP) and dynamic programming (DP) techniques are available in the literature. Among them Labye's method is especially well known in the design of irrigation systems (Labye, 1966, 1981). This approach initially assigns the minimum possible diameter to each link without infringing on the maximum velocity restriction. From this initial situation, it is based on the concept of economic slope β_s, which is defined as the quotient between the cost increase ($P_{s+1} - P_s$) produced when the diameter of the link is increased and the consequent gain in head loss ($J_s - J_{s+1}$). The economic slope enables the characteristic curve of each sub-network to be established. Each increase in diameter is decided while trying to minimize the associated cost increase in an iterative process that ends when all the pressure heads in the network coincide with the design head.

The linear programming method is also still accepted as an approach for the optimal selection of the diameters of pipes in branched irrigation networks. Mathematical models based on LP were initially developed and used in the irrigation network design process in the former Czechoslovakia in the early 1960s (Zdražil, 1965).

Various methods for optimizing branched irrigation networks - Labye's method, the linear programming method and a simplified nonlinear method - were compared in (Theocharis, et al., 2010).

The above-mentioned methodologies are suitable only for networks without loops, which are typical in the irrigation industry. However, there are frequent situations where the presence of loops in the network is useful, e.g., when redundant parallel pipes in the network's supply are needed or the interconnection of branches to correct shortages or equalize pressure solves some design problems. A possible approach for increasing the hydraulic capacity of branch systems in the rehabilitation process is to convert them to looped networks (thereby providing alternative pathways) and increase their hydraulic capacity with a minimum capital investment. In the majority of cases the requirement to increase the hydraulic capacity of the system could be based on the following requirements, e.g.:

- To increase pressure and water demands at hydrants due to upgrading the irrigators (with increased pressure and demand characteristics).
- To provide sufficient pressure within the pipeline system with an increased number of demand points as well as grouping them in selected parts of the irrigation system.
- To expand the system by adding new branches.
- To eliminate a system's deficiencies due to its aging.

The methodologies for optimizing looped water distribution systems (WDS) mainly evolved for drinking water distribution systems, but, with some modifications, they are also applied in the irrigation industry. Alperovits and Shamir (Alperovits & Shamir, 1977) extended the basic LP procedure to looped networks. Kessler and Shamir (Kessler & Shamir, 1989) used the linear programming gradient method as an extension of this method. It consists of two stages: an LP problem is solved for a given flow distribution, and then a search is conducted in the space of the flow variables. Later, Fujiwara and Khang (1990) used a two-phase decomposition method extending that of Alperovits and Shamir to non-linear modelling. Also, Eiger, et al. (1994) used the same formulation as Kessler and Shamir, which leads to a determination of the lengths of one or more segments in each link with discrete diameters. Nevertheless, these methods fail to resolve the problems of large looped systems.

Researchers have focused on stochastic or so-called heuristic optimization methods since the early 1990s. Simpson and his co-workers (1994) used basic genetic algorithms (GA). The simple GA was then improved by Dandy, et al. (1996) using the concept of the variable power scaling of the fitness function, an adjacency mutation operator, and gray codes. Savic and Walters (1997) also used a simple GA in conjunction with an EPANET network solver.

Other heuristic techniques have also been applied to the optimization of a looped water distribution system, such as simulated annealing (Loganathan, et al, 1995; Cunha and Sousa, 2001); an ant colony optimization algorithm (Maier, et al., 2003); a shuffled frog leaping algorithm (Eusuff & Lansey, 2003) and a harmony search (Geem, 2002), to name a few.

The impetus for this work is that significant differences from the known global optimums are referred to even for single objective tasks and simple benchmark networks, while existing algorithms are applied. Reca, et al. (2008) evaluated the performance of several meta-heuristic techniques - genetic algorithms, simulated annealing, tabu search and iterated local search. He compared these techniques by applying them to medium-sized benchmark networks. For the Hanoi network (which is a well-known benchmark often used in the optimization community), after ten different runs with five heuristic search techniques he obtained results which varied in a range from 6,173,421 to 6,352,526. These results differ by 1.5 - 4.5 % from the known global optimum for this task (6,081,128), which is a relatively large deviation for such a small network (it consists of 34 pipes). Similar results were presented by Zecchin, et al. (2007) and Cisty, et al. (1999).

The main concern of this paper is to propose a method which is more dependable and converges more closely to a global optimum than existing algorithms do. The paper proposes a new multiphase methodology for solving the optimal design of a water distribution system, based on a combination of differential evolution (DE) and particle swarm optimization (PSO) called DEPSO (Zhang & Xie, 2003). DEPSO has a consistently impressive performance in solving many real-world optimization problems (Xu, et al., 2007; Moore & Venayagamoorthy, 2006; Luitel & Venayagamoorthy, 2008; Xu, et al., 2010). As will be explained in the following text, the search process in PSO is based on social and cognitive components. The entire swarm tries to follow the global best solution, thus improving its own position. But for the particular particle that is the global best solution, the new velocity depends solely on the weighted old velocity. DEPSO adds the DE operator to the PSO procedure in order to add diversity to the PSO, thus keeping the particles from falling into a local minimum.

The second base improvement proposed, which should determine the effectiveness of the proposed methodology, is the application of a multi-step procedure together with the mentioned DEPSO methodology. The multi-step optimization procedure means that the optimization is accomplished in two or more phases (optimization runs) and that in each further run, the optimization problem comes with a reduced search space. This reduction of the search space is based on an assumption of the significant similarity between the flows in the sub-optimal solutions and the flows in the global optimal solution. The details are described later in this paper.

This chapter is structured as follows: In the "Methodology" section, WDS optimization is explained and formally defined. This section subsequently describes PSO, DE, and DEPSO, together with DEPSO's multi-phase application to WDS optimal design. The experimental data, design, and results are presented and discussed in the "Application and Results" section. The "Conclusion" section describes the main achievements.

2. Methodology

2.1 Optimal design of a water distribution network

Given a water network comprised of n nodes and l sizable components (pipes, valves, pumps and tanks), the general least-cost optimisation problem may be stated mathematically in terms of the various design variables x, nodal demands d, and nodal pressure heads h. Here x is a vector of the selected characteristic values (or physical dimensions) for the l sizable system components; d is a vector of length n specifying the demand flow rates at each node, and h is a vector of length n, whose entries are the pressure head values for the n nodes in the system (note that the head depends on x and d). Here x may include, for example, the diameter of the pipes, the capacity of the pumps, valve types and settings, and the tank volume, diameter and base elevation (Rossman, 2000). In our work the least cost optimal design problem is solved, and the decision variables are the diameters of the pipelines, which must be selected from a discrete set of commercially available pipe diameters.

The design constraints are typically determined by the minimal pressure head requirements at each demand node and the physical laws governing the flow dynamics. The objective is to minimize the cost function $f(h, d, x)$. This cost function may include installation costs, material costs, and the present value of the running costs and/or maintenance costs for a potential system over its entire lifetime. For optimisation methods that cannot explicitly accommodate constraints, it is a common practice to add a penalty term to the cost function, in order to penalize any constraint violations (such as deviation from the system's pressure requirements) (Lansey, 2000). This technique requires a penalty factor to scale the constraint violations to the same magnitude as the costs.

The WDS design optimization problem is therefore to

minimize

$$f(h, d, x),$$

subject to

$$g(q, d, x) = 0,$$

$$e(h, d, x) = 0,$$

$$h_{min} \le h(d, x) \le h_{max},$$

$$j_{min} \le j(x) \le j_{max} \tag{1}$$

where a set of at least n conservation of mass constraints $g(q, d, x) = 0$ includes the conservation of the flow equation for each of the nodes in the system, incorporating the nodal water demands d and the flows q for all the pipes branching from a node; the system of equations $e(q, d, x) = 0$ are energy equation constraints, specifying that energy is conserved around each loop, which then follows the pressure head constraints. The design constraints of the form $j_{min} < j(x) < j_{max}$ on the variables $j(x)$ specify the physical limitations or characteristic value sets from which the components may be selected (Lansey, 2000). These constraints may represent restrictions on discrete variables such as pipes which come in a range of commercial diameters.

The main design constraints (pressure head requirements) in the present work were determined by the EPANET 2 (Rossman, 2000) simulation model. For the purpose of the optimal design the model is first set up by incorporating all the options for the individual network components. The DEPSO then generates trial solutions, each of which is evaluated by simulating its hydraulic performance. Any hydraulic infeasibility, for example, failure to reach a specified minimum pressure at any demand point, is noted, and a penalty cost is calculated. The operational (e.g. energy) costs can also be calculated at this point if required. The penalty costs are then combined with the predicted capital and operational costs to obtain an overall measure of the quality of the trial solution. From this quality measure the fitness of the trial solution is derived. The process will continue for many thousands of iterations, and a population of good feasible solutions will evolve.

2.2 Particle swarm optimization

Particle swarm optimization (PSO) is a meta-heuristic method inspired by the flocking behaviour of animals and insect swarms. Kennedy and Eberhart (Kennedy et al., 2001) proposed the original PSO in 1995; since then it has steadily gained popularity. In PSO an individual solution in a population is treated as a particle flying through the search space, each of which is associated with a current velocity and memory of its previous best position, a knowledge of the global best position and, in some cases, a local best position within some neighbourhood - defined either in terms of the distance in decision/objective space or by some neighbourhood topology. The particles are initialized with a random velocity at a random starting position.

These components are represented in terms of the two best locations during the evolution process: one is the particle's own previous best position, recorded as vector p_i, according to the calculated fitness value, which is measured in terms of the clustering validity indices in the context of the clustering, and the other is the best position in the entire swarm, represented as p_g. Also, p_g can be replaced with a local best solution obtained within a certain local topological neighbourhood. The corresponding canonical PSO velocity and position equations at iteration t are written as

$$v_i(t) = w.v_i(t-1)+c_1.\varphi_1.(p_i - z_i(t-1)) + c_2.\varphi_2.(p_g - z_g(t-1)) \qquad (2)$$

$$z_i^t(t) = z_i\ (t-1)+v_i(t) \qquad (3)$$

where w is the inertial weight; c_1 and c_2 are the acceleration constants, and φ_1 and φ_2 are uniform random functions in the range of $[0,1]$. Parameters c_1 and c_2 are known as the cognitive and social components, respectively, and are used to adjust the velocity of a particle towards p_i and p_g.

PSO requires four user-dependent parameters, but accompanied by some useful rules. The inertia weight w is designed as a trade off between the global and local searches. The greater values of w facilitate global exploration, while the lower values encourage a local search. Parameter w can be a fixed to some certain value or can vary with a random component, such as:

$$w = w_{max} - \varphi_3/2, \qquad (4)$$

where φ_3 is a uniform random function in the range of $[0,1]$ and w_{max} is a constant. As an example, if w_{max} is set as 1, Eq. 4 makes w vary between 0.5 and 1, with a mean of 0.75. During the evolutionary procedure, the velocity for each particle is restricted to a limit w_{max}, as in velocity initialization. When the velocity exceeds w_{max}, it is reassigned to w_{max}. If w_{max} is too small, the particles may become trapped in the local optima, where if w_{max} is too large, the particles may miss some good solutions. Parameter w_{max} is usually set to around 10 - 20% of the dynamic range of the variable on each dimension (Kennedy et al., 2001).

Izquierdo et al. (2008) applied PSO to the water distribution system design optimization problem in his work. They developed an adaptation of the original algorithm, whereby the solution collisions (a problem that occurs frequently in PSO) are checked using several of the fittest particles, and any colliding solutions are randomly regenerated with a new position and velocity. This adaptation greatly improves the population diversity and global convergence characteristics. Finally, they adapted the algorithm to accommodate discrete variables by discretizing the velocities in order to create discrete step trajectories for these variables. Izquierdo et al. tested their algorithm on the NYTUN and HANOI WDS benchmarks and achieved large computational savings (an order of magnitude better than the previous methods), whilst closely approximating the known global optimum solutions.

2.3 Differential evolution

In 1997 Storn and Price (1997) first proposed differential evolution (DE), as a generic metaheuristic for the optimization of nonlinear and non-differentiable continuous space functions; it has proven to be very robust and competitive with respect to other evolutionary algorithms. At the heart of its success lies a very simple differential operator, whereby a trial solution vector is generated by mutating a random target vector by some multiple of the difference vector between two other random population members. For the three distinct random indices i, j and k, this has the form:

$$y_i = x_i + \hat{f} \times (x_j - x_k), \tag{5}$$

where x_i is the target vector; y_i is the trial vector; and $\hat{f}$ is a constant factor in the range $[0, 2]$ which controls the amplification of any differential variation, typically taken as 0.5. If the trial vector has a better objective function value, then it replaces its parent vector. Storn and Price also included a crossover operator between the trial vector and the target vector in order to improve convergence.

2.4 Hybrid DEPSO methodology

The DEPSO algorithm involves a two-step process. In the first step, the original PSO as previously described is applied. In the second step, the DE mutation operator is applied to the particles. The crossover rate for this study is given as one (Zhang & Xie, 2003). Therefore, for every odd iteration, the original PSO algorithm is carried out, while for every even iteration, the DEPSO algorithm is carried out. The procedure for the implementation of DEPSO is summarized in the following steps:

1. Initialize a population of particles with random positions and velocities. Set the values of the user-dependent parameters.
2. For every odd iteration, carry out the canonical PSO operation on each individual member of the population.
a. Calculate the fitness function $Fit(z_i)$ for each particle z_i;
b. Compare the fitness value of each particle $Fit(z_i)$ with $Fit(p_i)$. If the current value is better, reset both $Fit(pi)$ and p_i to the current value and location;
c. Compare the fitness value of each particle $Fit(z_i)$ with $Fit(p_g)$. If the current value is better, reset $Fit(p_g)$ and p_g to the current value and location;

d.   Update the velocity and position of the particles based on Eqs. 6 and 7.
3.   For every even iteration, carry out the following steps:
a.   For every particle z_i with its personal best p_i, randomly select four particles, z_a, z_b, z_c, and z_d, that are different from z_i and calculate Δ_1 and Δ_2 as,

$$\Delta_1 = p_a - p_b \, , a \neq b, \tag{6}$$

$$\Delta_2 = p_c - p_d \, , c \neq d, \tag{7}$$

where p_a, p_b, p_c, and p_d are the corresponding best solutions of the four selected particles.

b.   Calculate the mutation value δ_i by Eq. 8 and create the offspring o_{ij} by Eq.9,

$$\delta_i = (\Delta_1 + \Delta_2)/\ 2 \, , \tag{8}$$

$$o_{ij} = p_{ij} + \delta_{ij}, \text{ if } \varphi \leq p_r \text{ or } j = r \tag{9}$$

where j corresponds to the dimension of the individual, and r is a random integer within 1 and the dimension of the problem space.

c.   Once the new population of offspring is created using steps a) and b), their fitness is evaluated against that of the parent. The one with the higher fitness is selected to participate in the next generation.
d.   Recalculate the p_g and p_i of the new population.
4.   Repeat steps 2) to 3) until a stopping criterion is met, which usually occurs upon reaching the maximum number of iterations or discovering high-quality solutions.

2.5 Multi-step approach to WDS design

The proposed approach to the WDS optimization methodology involves refining the optimization calculations in a multiple-step approach, where the search space from the first optimization run is reduced for the second optimization run. In every run the DEPSO methodology is applied. The size of the search space depends on the number of possible diameters for each link from which the optimal option could be selected. In the first phase for all the links, all the available diameters are usually considered. In this case the size of the search space is n^l, where n is the number of possible diameters and l is the number of links. In the second phase the size of the search space is $n_1.n_2.n_3 \ldots n_l$, where n_i is the number of possible diameters for link i, which is a smaller number than in the first phase if some of the n_i are less than n.

On the basis of the flows computed in the pipes of the suboptimal solution in the first phase, it is possible, with the help of the known design minimum and maximum flow velocities, to calculate the maximal and minimal pipeline diameter considered for a given link of the WDS network. The prerequisite for undertaking such a step is the ability of modern heuristic algorithms to approximate the global optimum with a sufficient degree of accuracy, which could now, after two decades of their development, be expected. It is therefore assumed that the resulting suboptimal solution already has flows sufficiently close to the flows in the global optimum design of the WDS. This assumption is empirically verified by the author in this paper, but it also has a logical basis, since it is known that for a given distribution of the flows in a water distribution network, multiple solutions for the design of the diameters (the main design parameter in our definition of WDS optimization) could be found to comply with the technical requirements of the system.

One of the diameter designs for flow distribution in a network is best with regard to the cost of the network. This means that there are fewer variations in the flows than there are variations of the possible diameters, so if the diameters proposed by the heuristic search engine (e.g. DEPSO) differ from the optimal diameters searched for, the flows could be quite close to them, especially when a suboptimal solution close enough to the global optimal one is considered. There is some degree of intuition in this theorem, but the proposed idea was tested with positive results (as will be referred to hereinafter). The subsequent task is to find the corresponding optimal diameters for this distribution of the flows.

A reduction of the search space is accomplished for the second optimisation run with the assistance of the minimum and maximum pipeline flow velocities allowed. These parameters allow for the calculation of the anticipated minimum and maximum diameter for every network segment. These two values set an upper and lower boundary to the new range of acceptable diameters for each pipe segment, from which the algorithm will choose the optimal values in the second run. With the above-mentioned reduction of possible particle values a smaller search space is obtained, and better search results can be expected.

3. Application and results

The Tomasovo irrigation network was used as a case study in this work. Its layout is shown in Figure 1. This is one of the irrigation facilities in Slovakia which has medium-size area coverage, and the sprinkler type of irrigation is applied. Its construction was completed at the beginning of the 1960s; the whole facility is therefore approaching the end of its service life and can be selected as a suitable model for testing the proposed optimization methods which could also be easily applied for the rehabilitation of the hydraulic system.

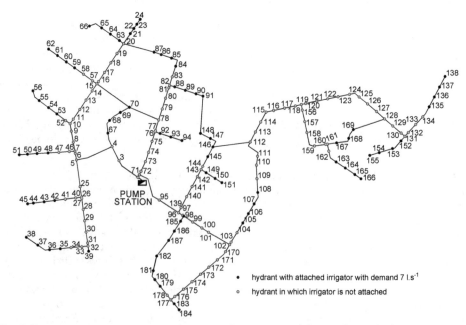

Fig. 1. Tomasovo irrigation system layout with positions of demands marked

The irrigated area of this system amounts to 700 ha. The hydraulic part of the irrigation system consists of the irrigation water take-off structure from the canal, a pump station, a pressurized network for the delivery of irrigation water, and sprinklers. For the purpose of this study it is necessary to describe only the pipeline network in detail (which is available from the author of this chapter in the form of an EPANET input file). The pump station and sprinklers represent the boundary conditions for the system analysed. Only their basic parameters (the pressure and output flow of the pump station, the required sprinkler pressure and demand flow) are taken into account in the optimization computations. A concept of irrigation with hose-reel irrigators with an optimum demand flow of 7.0 $l.s^{-1}$ and an optimal inlet pressure of 0.55 MPa is proposed for this system. In addition, it is assumed that a battery of such sprinklers will be used, i.e., there will be a set of four machines which operate as a whole on the adjacent hydrants – this approach has some advantages while managing the system, and similar operational rules could also be defined in other systems. Water is supplied through a pump station with an output pressure of 0.85 MPa and a flow of 392 $l.s^{-1}$. This means that 56 irrigators with a demand flow of 7.0 $l.s^{-1}$ could simultaneously work on the network (14 groups of irrigators). These irrigators could be placed in various hydrants of the network during the operation of this system. The worst case for their placement should be used for the design of the diameters. In Figure 1 the placement of the irrigators is displayed, the positions of which were used in searching for the optimal pipe diameters in this study. In some of the nodes the input of the water to the network is imitated in the EPANET inputs with the aim of reducing the overall flow to the maximal possible flow from the pump station and concurrently having in each part of the network such a flow which could be expected in this place during the system's operation.

This network has a total of 186 nodes supplied by one source node (the pump station). There are 193 pipes arranged in 7 loops, which are to be designed using a set of 7 asbestos cement pipes with diameters of 100, 150, 200, 300, 350, 400 and 500 mm and an absolute roughness coefficient of k = 0.05 mm. When searching for the optimal diameters for this system, a total enumeration of all possible alternatives from which the optimal solution should be chosen is not possible. The amount of possible combinations could be evaluated from a number of the above-mentioned proposed diameters powered on a number of pipes - it reaches the impressive amount of 7^{193}, which is a number with 163 digits before the decimal point. That is why the optimization methodologies described in the methodology part were applied for solving this task. The Darcy–Weisbach equation has been adapted to calculate the head losses, using EPANET 2 (Rossman, 2000). The minimum required pressure head in this network is 0.55 MPa for each demand node (which the proposed hose-reel sprinkler needs for its operation).

The computational experiments were accomplished in the following manner: Firstly, 100 testing runs of the first phase of the proposed algorithm (without reducing the search space) were computed for the Tomasovo network; the results are summarised in Figure 2. In this stage various DEPSO settings were used (DE factor=0.3÷0.8, CR=0.5÷1.0, PSO C1=C2=0.5÷1.5; $N_{population}$=100÷250 and $N_{generation}$= 500÷1000). The histogram in Figure 2 shows that the minimal (best) obtained cost of the optimized network in this phase was 510,469.5 €; the maximal (worst) network price was 544,464.8 €; and the most frequently obtained result was 515,000 € – 525,000 €. The original search space was reduced by the procedure explained in section 2.5 from the original 7^{193} to a value of 7^{100} on the basis of the flows in the most frequently obtained result (or average result) from this interval (the cost of this solution was 521,536.9 €). The mentioned one hundred runs of the first phase of the algorithm were performed with the intention of verifying the probability of obtaining this result (a reduction of the search space to approximately 7^{100} alternatives), which is a prerequisite for the next computational phase, e.g., this amount of the computations was accomplished only for testing purposes. In actual computations this is not necessary: five to ten runs would be enough, and the best solution in a real case could be taken as the basis for reducing the search space.

Thus in our testing computations, the reduced search space with an average reduction (which is also the most likely result obtained according to Figure 2) was chosen and entered into the second optimization run. The leading factor determining the search space reduction and affecting the accuracy of the calculations are, in addition, the mentioned result of the computations from the first phase of the algorithm and also the minimum and maximum flow velocities mentioned in section 2.5. The values of the velocity for the search reduction were 0.1 m.s^{-1} (v_{min}) and 3 m.s^{-1} (v_{max}). One hundred runs of the second phase were conducted similarly as in the case of the first phase computations in order to verify the probability of obtaining the final result, which is also reported in Figure 2. It is possible to see there that almost all the results from the second phase are on the left side of the results from the first phase, i.e., they are better. This means that it is better to apply our proposed two-phase algorithm than to refine or accomplish more computations without a reduction of the search space as is usual. The minimal (best) obtained cost of the optimized network in this final phase was 507,148.3 €; the maximal (worst) network cost was 513,462.0 €; and the average result was 508,970.5 €.

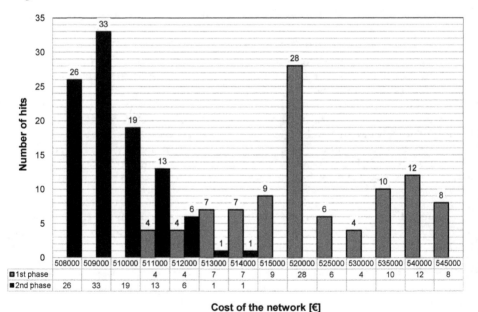

Fig. 2. Histogram of first and the second phases of the optimization computations

This procedure works fully automatically and does not need an expert's assistance in the optimization calculations. The EPANET input file of this water distribution network and all the results of the computations are not presented here in the table form in detail, because an inappropriately large space would be needed for such a presentation and is available from the author of this chapter.

3.1 A Comparison of the branched and looped alternatives of the irrigation network design

Irrigation systems were usually designed with branch layout. Because we proposed in this study procedure for design of the looped irrigation networks in this chapter are optimal designs for this two possibilities evaluated. The looped layout of the tested irrigation network is reduced to branch one by removing pipes between nodes 15-70, 20-87, 70-78, 91-148, 112-146, 128-169, 182-187. The linear programming (LP) method is accepted as an approach for the optimal selection of the diameters for pipes in branched networks. For the clarity purposes we briefly describe the optimisation procedure of the pipeline network rehabilitation using linear programming. The mathematical formulation of this problem is as follows:

$$A_{11}x_1 + A_{12}x_2 + ...+ A_{1n}x_n = B_1$$

$$A_{21}x_1 + A_{22}x_2 + ...+ A_{2n}x_n = B_2$$

etc.

$$A_{m1}x_1 + A_{m2}x_2 + ...+ A_{mn}x_n = B_m \tag{10}$$

$$c_1x_1 + c_2x_2 + ...+ c_3x_n = min \tag{11}$$

Solution has to comply with inequalities:

$$x_1 > 0; x_2 > 0 \text{ etc. up to } x_n > 0 \tag{12}$$

When in order to resolve pipeline networks optimization task linear programming is applied, unknown are the lengths of individual pipeline diameters. In conditions (10) should be mathematically expressed the requirement that the sum of unknown lengths of individual diameters in each section has to be equal to its total length. The second type of the equation in constraints (10) represents the request that the total pressure losses in a hydraulic path between the pump station and critical node (the end of the pipeline, extreme elevation inside the network) should be equal or less than the known value. This constraint is based on the maximum network pressure requirement needed for the operation of the system. Given the investment costs minimisation requirement, the objective function (11) sums the products of individual pipeline prices and their required lengths. Four possible diameters (base of v_{min} a v_{max}) are selected for each section. Further details on LP optimisation can be found in available literature, e.g., Cisty et al., (1999). The results of optimal design of the branch network by LP is summarised in the Table 1.

The results obtained indicate that the optimal design of a branched network using linear programming provides better results from an investment cost point of view (504,574.5 €)

than the calculations using DEPSO on a looped network (507,148.3 €). This follows from the fact that LP is a deterministic algorithm, which provides a real global minimum of the problem, which was defined by equations (10, 11, 12). The DEPSO method is a heuristic algorithm, which can provide results closer to a global minimum. The main reason is, of course, that there are fewer pipes in the branched alternative than in looped one. Considering the operation of an irrigation network, there are some advantages in using a looped layout, which is illustrated by evaluating the set of the real operation situations of the Tomasovo irrigation network both with branched and looped optimal designs.

The pressure assessment of the pipeline network was done in such a way that a set of realistic operational situations was analysed. These demand situations are proposed to have maximum hydraulic requirements (compatible with those used for the design of the network), and 320 various possibilities with different placements of the irrigators on the network were generated and evaluated. The next step was to run a simulation calculation of the branched and looped network configurations for all of these operational situations. In these alternatives we have assessed the minimal, maximal and average pressures at all the demand points. These values are shown in the diagram (Figure 3), where the data is sorted according to size.

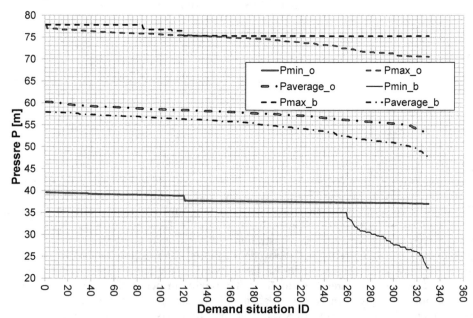

Fig. 3. Comparision of the pressures in 320 demand situations in the branched and looped alternatives

The simulation results prove the benefit of looping in hydraulic terms (better pressure ratios, lower maximal pressures, higher minimal pressures) and in economic terms – looped network rehabilitation is not much more expensive than a branched solution. There are unacceptably low pressures in branch networks in approximately 25% of the demand situations investigated, which is why a looped network should be preferred to a branched one from an operational point of view. One can assume that the results described are also applicable when designing other systems.

Diameter	unit cost	Branch network		Looped network first phase		Looped network second phase	
		length	cost	length	cost	length	cost
mm	€/m	m	€	m	€	m	€
100	15.5	5,049.6	78,268.8	8,679.7	134,535.35	8,582.2	133,024.1
150	20	7,509.8	150,196	5,671.1	113,422	6,302	126,040
200	12	6,629.8	79,557.6	8,523.9	102,286.8	8,273.7	99,284.4
300	36.5	3,756.4	137,108.6	3,194.7	116,606.55	3,015.5	110,065.7
350	47	1,171	55,037	834.3	39,212.1	730.3	34,324.1
400	55	35.3	1,941.5	35.3	1,941.5	35.3	1,941.5
500	72.5	34	2,465	34	2,465	34	2465
SUM		24,185.9	504,574.5	26,973	510,469.3	26,973	507,144.3

Table 1. Costs of the optimal design from linear programming and the first and second phase of the DEPSO algorithm

4. Conclusion

In this study the application of the DEPSO optimization algorithm for the design of a pressurised irrigation water distribution network is proposed. Its effectiveness is determined by the proposed multiple-step approach with application of the DEPSO heuristic methodology, where the optimized problem with a reduced search space is entered into each subsequent run. This reduction was obtained with the help of the assumption of a significant closeness between cost flows in the suboptimal and global optimal solutions. This assumption was empirically verified at the large Tomasovo irrigation network where this methodology was applied. The calculation results for this network show the better performance of the proposed methodology compared to the traditional, one-step application of the various heuristic methods. The benefit of designing the looped alternative versus the branch one is demonstrated by comparing the operational flexibility of networks designed by DEPSO and by linear programming.

The focus of the work was aimed at simplifying the calculations for practical use. The proposed optimization procedure could work fully automatically and does not need an expert's assistance in the optimization calculations (e.g., for choosing the various parameters of the heuristic methodology). Various improvements are possible in future research, e.g., the direct inclusion of the operation evaluation into the optimization procedure by applying a multi-objective approach.

5. Acknowledgment

This work was supported by the Slovak Research and Development Agency under Contract No. LPP-0319-09, and by the Scientific Grant Agency of the Ministry of Education of the Slovak Republic and the Slovak Academy of Sciences, Grant No. 1/1044/11.

6. References

Alperovits, E. & Shamir U. (1977). Design of optimal water distribution systems. *Water Resource Research,*Vol. 13, No.6, pp. 885-900

Cisty, M., Savic, D.A., Walters, G.A. (1999). Rehabilitation of pressurized pipe networks using genetic algorithms. *Proceedings Water for Agriculture in the Next Millennium. 17th Congress on Irrigation and Drainage*, International Commission on Irrigation and Drainage, Granada, pp. 13-27.

Cunha, M.C. & Sousa, J. (2001). Hydraulic infrastructures design using simulated annealing. *Journal of Infrastructure Systems*, ASCE, Vol.7, No.1, pp. 32-39

Dandy, G.C., Simpson, A.R. & Murphy, L.J. (1996). An Improved Genetic Algorithm for Pipe Network Optimization. *Water Resources Research*, Vol.32, No.2, pp. 449-458

Eiger, G. , Shamir, U. & Ben-Tal, A. (1994). Optimal design of water distribution networks. *Water Resources Research*, Vol.30, No.9, pp. 2637-2646

Eusuff, M.M. & Lansey, K.E. (2003). Optimization of water distribution network design using the shuffled frog leaping algorithm. *Journal of Water Resources Planning and Management ASCE*, Vol.129, No.3, pp. 210-225

Fujiwara, O. & Khang, D.B. (1990). A Two-Phase Decomposition Method for Optimal Design of Looped Water Distribution Networks. *Water Resources Research*, Vol.26, No.4, pp. 539-549

Geem, Z.W., Kim J.H. & Loganathan, G.V. (2003). Harmony search optimization: application to pipe network design. *International Journal of Modelling and Simulation*, Vol.22, No.2, pp. 125-133

González-Cebollada, C., Macarulla, B. & Sallán D. (2011). Recursive Design of Pressurized Branched Irrigation Networks. *Journal of Irrigation and Drainage Engineering*, Vol. 137, No. 6, June 1, pp. 375 – 382

Izquierdo, J., Montalvo, I., Peacuterez R. & Iglesias, P.L. (2008). A diversity-enriched variant of discrete PSO applied to the design of water distribution networks. *Engineering Optimization*, Vol.40, No.7, pp. 655-668

Kennedy, J., Eberhart, R. & Shi, Y. (2001). *Swarm Intelligence*. San Diego, CA: Academic Press

Kessler, A. & Shamir, U. (1989). Analysis of the Linear Programming Gradient Method for Optimal Design of Water Supply Networks. *Water Resources Research*, Vol.25, No.7, pp. 1469-1480

Labye, Y. (1966). Etude des procedés de calcul ayant pour but de rendre minimal le cout d'un reseau de distribution d'eau sous pression, La Houille Blanche 5, pp. 577–583

Lansey, K.E. (2000). Optimal Design of Water Distribution Systems, In "Water Distribution Systems Handbook", Mays, L.W (editor), McGraw-Hill, New York,.

Loganathan, G.V., Greene J.J. & Ahn, T.J. (1995). Design heuristic for globally minimum cost water-distribution systems. *Journal of Water Resources Planning and Management,* Vol.121, No.2, pp. 182–192

Luitel, B. & Venayagamoorthy, G. (2008). Differential evolution particle swarm optimization for digital filter design. *Proceedings of IEEE Congress on Evolutionary Computation,* Hong Kong, China, June 1-6, 3954–3961

Maier, H.R., Simpson, A.R., Foong, W.K., Phang, K.Y., Seah, H.Y., & Tan, C.L. (2001). Ant colony optimization for the design of water distribution systems. *Proceedings of the World Water and Environmental Resources Congress,* Orlando, Florida, May 20-24

Moore, P. & Venayagamoorthy, G. (2006). Evolving digital circuits using hybrid particle swarm optimization and differential evolution. *International Journal of Neural Systems,* Vol.16, No.3, pp. 163–177

Reca, J., Martínez, J., Gil, C., Baños, R. (2008). Application of several meta-heuristic techniquess to the optimization of real looped water distribution networks. *Water Resources Management,* Vol.22, No.10, pp. 1367–1379

Rossman, L. A (2000). *EPANET User's Manual,* U.S. Environmental Protection Agency

Savic, D.A. & Walters, G.A. (1997). Genetic algorithms for least-cost design of water distribution networks. *Journal of Water Resources Planning and Management,* ASCE, Vol. 123, No.2, pp. 67-77

Simpson, A.R., Dandy, G.C., Murphy, L.J. (1994). Genetic algorithms compared to other techniques for pipe optimization. *Journal of Water Resources Planning and Management,* ASCE, Vol.120, No.4, (July/August)

Storn, R. & Price, K. (1997). Differential evolution: A simple and efficient heuristic for global optimization over continuous spaces. *Journal of Global Optimization,* Vol.11, No.4, pp. 341-359,.

Theocharis, M.E., Tzimopoulos, C.D., Sakellariou-Makrantonaki, M.A., Yannopoulos, S.I., Meletiou, I.K. (2005). Comparative calculation of irrigation networks using Labye's method, the linear programming method and a simplified nonlinear method, *Mathematical and Computer Modelling,* Vol.51, No.3-4, pp. 286-299, ISSN 0895-7177

Xu, R., Venayagamoorthy, G. & Wunsch II, D. (2007). Modeling of gene regulatory networks with hybrid differential evolution and particle swarm optimization. *Neural Networks,* Vol. 20, No.8, pp. 917–927

Xu, R., Xu, J., Wunsch, D.C., (2010). Clustering with differential evolution particle swarm optimization. Evolutionary Computation (CEC), 2010 IEEE World Congress on Comutational Inteligence, Barcelona, Spain, July 18-23

Zdražil K.1986. Research of the computation and graphical methods of designing the irrigation pipe networks. VUZH Bratislava. In Czech.

Zecchin, A.C., Maier, H.R., Simpson, A.R., Leonard, M. & Nixon, J.B. (2007). Ant colony optimization applied to water distribution system design: comparative study of five algorithms. *Journal of Water Resources Planning and Management,* Vol. 133, No.1, pp. 87-92

Zhang W. & Xie, X. (2003). DEPSO: Hybrid particle swarm with differential evolution
operator. *Proceedings of IEEE International Conference on Systems, Man, and
Cybernetics*, Washington D C, USA, 2003, pp. 3816–3821

7

An Algebraic Approach for Controlling Cascade of Reaches in Irrigation Canals

Mouhamadou Samsidy Goudiaby[1], Abdou Sene[1] and Gunilla Kreiss[2]
[1]*LANI, UFR SAT, Université Gaston Berger, Saint-Louis*
[2]*Division of Scientific Computing, Department of Information Technology,*
Uppsala University, Uppsala
[1]*Senegal*
[2]*Sweden*

1. Introduction

Due to the lack of water resources, the problem of water management and minimizing the losses becomes an attraction for many researchers. Although some problems have been already solved in the theoritical point of view, only few of the proposed solutions have been effectively tested in a real situation (Litrico et al., 2003). Limitations of water control technology have been discussed in (Gowing., 1999). However, there are problems that have not been solved yet, as reported in (Bastin et al., 2009). Those problems concern both technological applications and mathematical challenges. To solve water management problems, the so-called St. Venant equations (De Saint-Venant., 1871) are often used as a fondamental tool to describe the dynamics of canals and rivers. They are composed by a 2×2 system of hyperbolic partial differential equations.

For a long time, the matter of controlling water level and flow in open canals has been considered in the literature. Various methods have been used to design boundary controllers which satisfy farmers or navigability demands. Among those different methods, we have: LQ (Linear Quadratic) control that has been particularly developped and studied by (Balogun et al., 1988), and (Malaterre., 1998) (see also (Weyer., 2003) and (Chen et al., 2002)). (Weyer., 2003) has considered LQ control of an irrigation canal in which the water levels are controlled using overshot gates located along the canal. A LQ control problem for linear symmetric infinite-dimensional systems has been considered by (Chen et al., 2002). PI (proportional and integral) control method has been used by (Xu & Sallet., 1999) to propose an output feedback controller using a linear PDE model around a steady state. Such an approach has been considered by (Litrico et al., 2003), where the authors expose and validate a methodology to design efficient automatic controllers for irrigation canals. Riemann and Lyapunov approaches are also considered (Leugering & Schmidt., 2002), (De Halleux et al., 2003), and recently by (Cen & Xi., 2009) and (Bastin et al., 2009).

For networks of open canals, many results have been shown by researchers using some of the methods mentionned above. For example, (De Halleux et al., 2003) have used the Riemann approach to deduce a stabilization control, for a network made up by several interconnected reaches in cascade (also (Cen & Xi., 2009) and reference cited therein). (Bastin et al., 2009) have used the Lyapunov stability approach to study the exponential stability (in L2-norm) of the classical solutions of the linearised Saint-Venant equations for the same network with a sloping bottom. (Leugering & Schmidt., 2002) have studied stabilization and null controlability of pertubations around a steady state for a star configuration network. Star configuration network can also be found in (Li., 2005) and (Goudiaby et al., -). (Goudiaby et al., -), have used a new approach to design boundary feedback controllers which stabilize the water flow and level around a given steady state.

Concerning network made up by several interconnected reaches in cascade, we have noticed, in the theoritical point of view, two approaches that are the Riemann invariants (De Halleux et al., 2003) and Lyapunov Analysis approaches (Bastin et al., 2009), (Cen & Xi., 2009). The purpose of this paper is to apply the approach given in (Goudiaby et al., -) to that network. The approach is applied to a network of two reaches but it can be generalize. Choosing a different type of network requires different treatment of junction where canals met together. On the other hand, the Saint-Venant equations considered in the present paper are in the non-conservation form. We consider the velocity at the boundaries as the controllable quantities.

The approach consists in expressing the rate of change of energy of the linearized problem, as a second order polynomial in terms of the flow velocity at the boundaries. The polynomial is handled in such a way to construct boundary feedback controllers that result in the water flow and the height approaching a given steady state. The water levels at the boundaries and at the junction are used to build the controllers. After deriving the controllers, we numerically apply them to a real problem, which is nonlinear, in order to investigate the robustness and flexibility of the approach.

The paper is organized as follows. In section 2, we present the network and the equations. We discuss how to determine a steady state solution and derive the linearized system and corresponding characteristic variables, on which controllers are built. We also formulate the main result, stating controllers and corresponding energy decay rates. In section 3, we demonstrate the approach by proving a corresponding result for a single reach, while the case of the network is proven in section 4. Numerical results obtained by a high order finite volume method (Leveque., 2002; Toro., 1999) are presented in section 5.

2. Governing equations and main result

The network can be given by Figure 1 or by any type of network where several reaches are interconnected in cascade (see (Bastin et al., 2009; De Halleux et al., 2003)). In Figure 1, M is considered as the junction node. The network model is given by the 1D St. Venant equations in each reach ($i = 1, 2$) and a flow conservation condition at M. The following variables are used: h_i is the height of the fluid column (m), v_i is the flow velocity (ms^{-1}), L_i is the length of the reach (m). The one dimensional St. Venant equations considered in the present paper are the following:

$$\begin{cases} \dfrac{\partial h_i}{\partial t} + \dfrac{\partial(v_i h_i)}{\partial x} = 0, & \text{in } [0, L_i] \\[4mm] \dfrac{\partial v_i}{\partial t} + \dfrac{1}{2}\dfrac{\partial v_i^2}{\partial x} + g\dfrac{\partial h_i}{\partial x} = 0, & \text{in } [0, L_i] \end{cases} \tag{1}$$

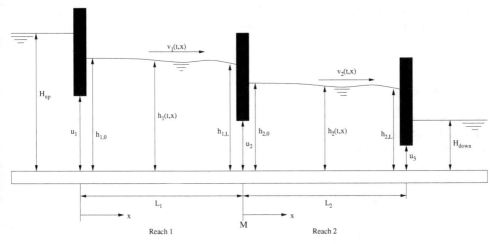

Fig. 1. The cascade network

together with a flow conservation condition at M,

$$h_1(t, L_1)v_1(t, L_1) = h_2(t, 0)v_2(t, 0), \tag{2}$$

initial conditions

$$h_i(0, x) = h_i^0(x), \quad v_i(0, x) = v_i^0(x), \tag{3}$$

and boundary conditions

$$v_1(t, 0) = v_{1,0}(t), \quad v_1(t, L_1) = v_{1,L_1}(t), \quad v_2(t, L_2) = v_{2,L_2}(t). \tag{4}$$

The results in the present paper concern a linearized system around a desired steady state. The controllers are built using that linear system and will be applied numerically to the above nonlinear model.

2.1 Steady state

The goal is to achieve a prescribed steady state $(\bar{h}_i, \bar{v}_i)$, with the help of the controllers, when time goes to infinity. From (1), the steady state solution $(\bar{h}_i, \bar{v}_i)$ satisfies:

$$\begin{cases} \dfrac{\partial \bar{v}_i}{\partial x} = 0, & \text{in} \quad [0, L_i], \\[3mm] \dfrac{\partial \bar{h}_i}{\partial x} = 0, & \text{in} \quad [0, L_i]. \\[3mm] \bar{h}_1(L_1)\bar{v}_1(L_1) = \bar{h}_2(0)\bar{v}_2(0) & \text{at} \quad M. \end{cases} \tag{5}$$

The steady state is such that

$$\bar{h}_2 < \bar{h}_1. \tag{6}$$

To determine the stady state (5), one gives $\bar{h}_1$, $\bar{v}_1$ and $\bar{h}_2$. On the other hand, using the flow direction (Figure 1) and the subcritical flow condition, one has

$$\bar{v}_i \geq 0 \quad \text{and} \quad \sqrt{g\bar{h}_i} > \bar{v}_i. \tag{7}$$

2.2 Linearized model

We introduce the residual state $(\check{h}_i, \check{v}_i)$ as the difference between the present state (h_i, v_i) and the steady state $(\bar{h}_i, \bar{v}_i)$: $\check{h}_i(t,x) = h_i(t,x) - \bar{h}_i(x)$, $\check{v}_i(t,x) = v_i(t,x) - \bar{v}_i(x)$. We use the assumptions $|\check{h}_i| \ll \bar{h}_i$ and $|\check{v}_i| \ll |\bar{v}_i|$ to linearize (1)-(4). Therefore, the solution $(\check{h}_i, \check{v}_i)$ satisfies

$$\begin{cases} (a) \; \dfrac{\partial \check{h}_i}{\partial t} + \bar{h}\dfrac{\partial \check{v}_i}{\partial x} + \bar{v}_i\dfrac{\partial \check{h}_i}{\partial x} = 0, \\[3mm] (b) \; \dfrac{\partial \check{v}_i}{\partial t} + \bar{v}_i\dfrac{\partial \check{v}_i}{\partial x} + g\dfrac{\partial \check{h}_i}{\partial x} = 0, \\[3mm] (c) \; \bar{v}_1\check{h}_1(t,L_1) + \bar{h}_1\check{v}_1(t,L_1) = \bar{v}_2\check{h}_2(t,0) + \bar{h}_2\check{v}_2(t,0) \quad \text{at} \quad M \end{cases} \tag{8}$$

together with the initial condition

$$\check{h}_i(0,x) = \check{h}_i^0(x), \quad \check{v}_i(0,x) = \check{v}_i^0(x), \tag{9}$$

and the boundary conditions as control laws

$$\check{v}_1(t,0) = \check{v}_{1,0}(t), \quad \check{v}_1(t,L_1) = \check{v}_{1,L_1}(t), \quad \check{v}_2(t,L_2) = \check{v}_{2,L_2}(t). \tag{10}$$

The functions $\check{v}_{1,0}(t)$, $\check{v}_{1,L_1}(t)$ and $\check{v}_{2,L_2}(t)$ are the feedback control laws to be prescribed in such a way to get an exponential convergence of $(\check{h}_i, \check{v}_i)$ to zero in time.

2.3 Eigenstructure and characteristic variables

The following characteristic variables are used to build the controllers:

$$\xi_{i1} = \check{v}_i - \check{h}_i\sqrt{\dfrac{g}{\bar{h}_i}} \quad \text{and} \quad \xi_{i2} = \check{v}_i + \check{h}_i\sqrt{\dfrac{g}{\bar{h}_i}}. \tag{11}$$

The characteristic velocities are

$$\lambda_{i1} = \bar{v}_i - \sqrt{g\bar{h}_i} \quad \text{and} \quad \lambda_{i2} = \bar{v}_i + \sqrt{g\bar{h}_i}.$$

The subcritical flow condition and the flow direction give

$$\lambda_{i1} < 0 < \lambda_{i2} \quad \text{and} \quad \lambda_{i1} + \lambda_{i2} \geq 0, \tag{12}$$

respectively. The characteristic variables satisfy

$$\frac{d\xi_{ij}}{dt} = \frac{\partial \xi_{ij}}{\partial t} + \lambda_{ij}\frac{\partial \xi_{ij}}{\partial x} = 0, \quad i,j = 1,2. \tag{13}$$

2.4 Main result

To build the feedback controllers, we express outgoing characteristic variables at the free endpoints and at the junction M in terms of initial data and the solution at the endpoints and at the junction M at earlier times. For reach 1, the outgoing characteristic variable at the endpoint $x = 0$ is ξ_{11}. For reach 2, the outgoing characteristic variable at the endpoint $x = L_2$ is ξ_{22}. Concerning the junction M, ξ_{12} and ξ_{21} are the outgoing characteristic variables. In section 4, we will see that

$$\begin{pmatrix} \xi_{11}(t,0) \\ \xi_{22}(t,L_2) \\ \xi_{12}(t,L_1) \\ \xi_{21}(t,0) \end{pmatrix} = \begin{pmatrix} b_1(t) \\ b_2(t) \\ b_3(t) \\ b_4(t) \end{pmatrix}, \tag{14}$$

where $b_i, i = 1, 2, 3, 4$ depend only on the initial condition and the solution at the endpoints and at the junction M at earlier times $\tau = t - \delta t$ with $\delta t \geq \min\left(\frac{L_1}{\lambda_{12}}, \frac{L_2}{\lambda_{22}}\right)$.

Let us consider $\theta_1 : \mathbb{R}^+ \longrightarrow]0,1]$ satisfying:

$$\theta_1(t) \geq \frac{2\bar{v}_1}{\lambda_{12}}. \tag{15}$$

and θ_2, $\theta_3 : \mathbb{R}^+ \longrightarrow]0,1]$ two arbitrary functions. We choose the feedback controllers as follows:

$$\check{v}_{1,0}(t) = -\frac{b_1(t)}{2}\left(\sqrt{1 - \theta_1(t)} - 1\right),$$

$$\check{v}_{2,L_2}(t) = -\frac{b_2(t)}{2}\left(\sqrt{1 - \theta_2(t)} - 1\right), \tag{16}$$

$$\check{v}_{1,L_1}(t) = \frac{\gamma(t)}{2\sigma}\left(\sqrt{1 - \theta_3(t)} - 1\right),$$

where,

$$\sigma = \bar{h}_1|\lambda_{11}|\left(1 + \frac{|\lambda_{11}|}{\lambda_{22}}\right), \qquad \gamma(t) = \bar{h}_1|\lambda_{11}|\left(1 - \frac{2\bar{v}_1}{\lambda_{22}}\right)b_3(t) + |\lambda_{11}|\sqrt{\bar{h}_1\bar{h}_2}\left(1 - \frac{2\bar{v}_2}{\lambda_{22}}\right)b_4(t),$$

and b_i, $i = 1, 2, 3, 4$, are given by (14). Therefore, defining

$$T = \max\left(\frac{L_1}{|\lambda_{11}|}, \frac{L_2}{|\lambda_{21}|}\right), \tag{17}$$

and the energy of the network by

$$E = \sum_{i=1}^{2} E_i, \quad E_i = \int_0^{L_i}\left(g\check{h}_i^2(t) + \bar{h}_i\check{v}_i^2(t)\right)dx, \tag{18}$$

we get the main result of this paper:

Theorem 1. *Let* $t_k = kT$, $k \in \mathbb{N}$, *where T is given by (17). Assume that the flow in the network is subcritical, the initial condition $(\mathring{h}_i^0, \mathring{q}_i^0)$ is continuous in $]0, L_i[$, $\check{v}_{1,0}, \check{v}_{1,L_1}, \check{v}_{2,L_2}$ satisfy (16), θ_1 satisfies (15) and $\lambda_{i1} + \lambda_{i2} \geq 0$. Then (8)-(10) has a unique solution $(\check{h}_i, \check{q}_i)$ continuous in $[t_k, t_{k+1}] \times]0, L_i[$ satisfying the following energy estimate:*

$$E(t_{k+1}) \leq (1 - \Theta^k) E(t_k), \tag{19}$$

where E is given by (18) and

$$\Theta^k = \min\left(\Gamma_1^k, \Gamma_2^k\right) \in [0, 1[,$$

$$\Gamma_1^k = \min\left(\inf_{x \in]0, L_1[}\left(\theta_1\left(t_k + \frac{x}{|\lambda_{11}|}\right) - \frac{2\bar{v}_1}{\lambda_{12}}\right), 4\bar{v}_1 \frac{(\bar{v}_2 - \bar{v}_1)}{\lambda_{22}\lambda_{12}}\right),$$

$$\Gamma_2^k = \min\left(\inf_{x \in]0, L_2[}\left(\frac{|\lambda_{21}|}{\lambda_{22}}\theta_2\left(t_k + \frac{L_2 - x}{\lambda_{22}}\right) + \frac{2\bar{v}_2}{\lambda_{22}}\right), 2\frac{(\bar{v}_2 - \bar{v}_1)}{\lambda_{22}}\left(1 - \frac{2\bar{v}_2}{\lambda_{22}}\right)\right).$$

Remark 1.

1. *In addition to (19), within the interval $]t_k, t_{k+1}[$, the energy is non-increasing.*
2. *The controllers (16) tend to zero when time goes to infinity. This is due to (19) and the fact that they are built on the solution at earlier times.*
3. *Estimation (19) can be written as*

$$E(t_k) \leq E(0) \exp\left(-\mu^k t_k\right).$$

where $\mu^k = \dfrac{1}{k}\displaystyle\sum_{j=0}^{k-1} \nu^j$ and $\nu^j = -\ln\left((1 - \Theta^j)^{\frac{1}{t_1}}\right)$. Thus, the functions θ can be viewed as stabilization rate for the exponential decrease.

3. Building the controller for a single reach

We construct a stabilization process for a single canal, which should drive the perturbations $\check{h}$ and $\check{v}$ to zero exponentially in time. We consider the 1D Saint-Venant equations (1) without the index i standing for the reach number:

$$\begin{cases} \dfrac{\partial h}{\partial t} + \dfrac{\partial(vh)}{\partial x} = 0, \\[3mm] \dfrac{\partial v}{\partial t} + \dfrac{1}{2}\dfrac{\partial v^2}{\partial x} + g\dfrac{\partial h}{\partial x} = 0, \end{cases} \tag{20}$$

together with initial conditions

$$h(0, x) = h^0(x), \quad v(0, x) = v^0(x) \tag{21}$$

and boundary conditions

$$v(t, 0) = v_0(t), \quad v(t, L) = v_L(t). \tag{22}$$

The steady state solution $(\bar{h}, \bar{v})$ satisfies:

$$\frac{\partial \bar{v}}{\partial x} = 0, \quad \frac{\partial \bar{h}}{\partial x} = 0, \quad \text{in} \quad [0, L].$$

with

$$\bar{v} \geq 0 \quad \text{and} \quad \sqrt{g\bar{h}} > \bar{v}. \tag{23}$$

The linearized model is

$$\begin{cases} (a) \ \dfrac{\partial \check{h}}{\partial t} + \bar{h}\dfrac{\partial \check{v}}{\partial x} + \bar{v}\dfrac{\partial \check{h}}{\partial x} = 0, \\[3mm] (b) \ \dfrac{\partial \check{v}}{\partial t} + \bar{v}\dfrac{\partial \check{v}}{\partial x} + g\dfrac{\partial \check{h}}{\partial x} = 0, \end{cases} \tag{24}$$

together with initial conditions

$$\check{h}(0, x) = \check{h}^0(x), \quad \check{v}(0, x) = \check{v}^0(x), \tag{25}$$

and the boundary conditions

$$\check{v}(t, 0) = \check{v}_0(t), \quad \check{v}(t, L) = \check{v}_L(t). \tag{26}$$

The functions $\check{v}_L(t)$ and $\check{v}_0(t)$ are the feedback control laws to be prescribed in such a way to get an exponential convergence of $(\check{h}, \check{v})$ to zero in time.

The characteristic variables are:

$$\zeta_1 = \check{v} - \check{h}\sqrt{\frac{g}{\bar{h}}} \quad \text{and} \quad \zeta_2 = \check{v} + \check{h}\sqrt{\frac{g}{\bar{h}}}, \tag{27}$$

with the characteristic velocities

$$\lambda_1 = \bar{v} - \sqrt{g\bar{h}} \quad \text{and} \quad \lambda_2 = \bar{v} + \sqrt{g\bar{h}}.$$

The subcritical flow condition and the flow direction give

$$\lambda_1 < 0 < \lambda_2 \quad \text{and} \quad \lambda_1 + \lambda_2 \geq 0, \tag{28}$$

respectively. Considering the characteristic variables (27), system (24) is written as two independant equations:

$$\frac{d\zeta_j}{dt} = \frac{\partial \zeta_j}{\partial t} + \lambda_j \frac{\partial \zeta_j}{\partial x} = 0, \quad j = 1, 2. \tag{29}$$

3.1 A priori energy estimation

Let E be the energy of (24) defined as

$$E(t) = \int_0^L \left(g\check{h}^2(t) + \bar{h}\check{v}^2(t) \right) dx. \tag{30}$$

We consider the following system as a weak formulation of (24)

$$\begin{cases} \forall (\psi, \phi) \in H^1(]0, L[), \\[2mm] \displaystyle\int_0^L g\psi \frac{\partial \check{h}}{\partial t}\, dx - g\bar{h} \int_0^L \check{v}\frac{\partial(\psi)}{\partial x}\, dx - g\bar{v} \int_0^L \check{h}\frac{\partial(\psi)}{\partial x}\, dx + \\[3mm] \quad g\bar{h}\psi(L)\check{v}_L(t) - g\bar{h}\psi(0)\check{v}_0(t) + g\bar{v}\psi(L)\check{h}_L(t) - g\bar{v}\psi(0)\check{h}_0(t) = 0, \qquad (31) \\[3mm] \displaystyle\int_0^L \bar{h}\phi \frac{\partial \check{v}}{\partial t}\, dx - \bar{h}\bar{v} \int_0^L \check{v}\frac{\partial(\phi)}{\partial x}\, dx - g\bar{h} \int_0^L \check{h}\frac{\partial(\phi)}{\partial x}\, dx \\[3mm] \quad + \bar{h}\bar{v}\phi(L)\check{v}_L(t) - \bar{h}\bar{v}\phi(0)\check{v}_0(t) + g\bar{h}\phi(L)\check{h}_L(t) - g\bar{h}\phi(0)\check{h}_0(t) = 0, \end{cases}$$

together with boundary and initial conditions.

We estimate the variation of the energy E on the canal in order to define the controllers $\check{v}_L(t)$ on $\{x = L\}$ and $\check{v}_0(t)$ on $\{x = 0\}$. To this end, we let $(\psi, \phi) = (\check{h}, \check{v})$ in (31) to get

$$\frac{1}{2}\frac{d}{dt}E(t) = -\frac{\bar{h}\bar{v}}{2}\check{v}_L^2(t) - \frac{g\bar{v}}{2}\check{h}^2(t, L) - g\bar{h}\check{h}(t, L)\check{v}_L(t)$$
$$+ \frac{\bar{h}\bar{v}}{2}\check{v}_0^2(t) + \frac{g\bar{v}}{2}\check{h}^2(t, 0) + g\bar{h}\check{h}(t, 0)\check{v}_0(t). \qquad (32)$$

The difference among control methods depends on how the energy is defined and its variation handled to obtain a convergence of the perturbations $\check{h}$ and $\check{v}$ to zero in time (see (Bastin et al., 2009; De Halleux et al., 2003)).

3.2 Controllers and the stabilization process

The feedback control building relies on the fact that we can express the height at the boundaries in terms of the flow velocity and outgoing characteristic variables. Using (29)

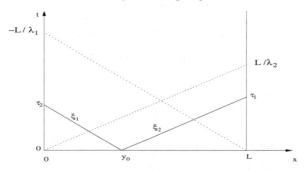

Fig. 2. Characteristic variables.

and refering on the characteristic variables indicated in figure 2, one has:

$$\begin{pmatrix} \xi_1(\tau_2, 0) \\ \xi_2(\tau_1, L) \end{pmatrix} = \begin{pmatrix} b_1(\tau_2) \\ b_2(\tau_1) \end{pmatrix}, \qquad (33)$$

where

$$b_1(\tau_2) = \begin{cases} \xi_1(0, |\lambda_1|\tau_2), & \tau_2 \le \frac{L}{|\lambda_1|}, \\ \xi_1(\tau_2 - \frac{L}{|\lambda_1|}, L), & \tau_2 \ge \frac{L}{|\lambda_1|}, \end{cases} \quad b_2(\tau_1) = \begin{cases} \xi_2(0, L - \lambda_2\tau_1), & \tau_1 \le \frac{L}{\lambda_2}, \\ \xi_2(\tau_1 - \frac{L}{\lambda_2}, 0), & \tau_1 \ge \frac{L}{\lambda_2}. \end{cases} \tag{34}$$

From (27), one derives

$$\check{h}(\tau_1, L) = \left(\xi_2(\tau_1, L) - v_L(\tau_1) \right) \sqrt{\frac{\bar{h}}{g}}, \tag{35}$$

$$\check{h}(\tau_2, 0) = \left(-\xi_1(\tau_2, 0) + v_0(\tau_2) \right) \sqrt{\frac{\bar{h}}{g}}. \tag{36}$$

Considering the energy equation (32), one deduces from (34)-(36) that

$$\frac{1}{2} \frac{dE}{dt}(t) = a_1 \check{v}_0^2(t) - a_1 b_1(t) \check{v}_0(t) + c_1(t) + a_2 \check{v}_L^2(t) - a_2 b_2(t) \check{v}_L(t) + c_2(t) \tag{37}$$

where

$$a_1 = \bar{h}\lambda_2, \quad a_2 = \bar{h}|\lambda_1|, \quad c_1(t) = \frac{\bar{h}\bar{v}}{2} b_1^2(t), \quad c_2(t) = -\frac{\bar{h}\bar{v}}{2} b_2^2(t), \tag{38}$$

$b_1(t)$ and $b_2(t)$ are given by (34).

The RHS of (37) is treated in such a way to get an exponential decrease of the energy. For this propose, the following observation for second order polynomials is used.

Lemma 1. *Consider a second order polynomial $P(q) = av^2 + bv$, where $a > 0$. For any $\theta \in [0, 1]$*

$$P\left(\frac{b}{2a}(\sqrt{1-\theta} - 1) \right) = -\frac{b^2}{4a}\theta. \tag{39}$$

If the flow velocity at the boundary is prescribed as follows:

$$\check{v}_L(t) = -\frac{b_2(t)}{2}\left(\sqrt{1 - \theta_2(t)} - 1 \right) \quad \text{and} \quad \check{v}_0(t) = -\frac{b_1(t)}{2}\left(\sqrt{1 - \theta_1(t)} - 1 \right), \tag{40}$$

where $\theta_1, \theta_2 : \mathbb{R}^+ \longrightarrow [0, 1]$, then by Lemma 1, (37) becomes

$$\frac{1}{2}\frac{dE}{dt}(t) = -\frac{b_1^2(t)}{4a_1}\theta_1(t) + c_1 - \frac{b_2^2}{4a_2}\theta_2(t) + c_2,$$

$$= -\frac{\bar{h}}{4}(\lambda_2\theta_1(t) - 2\bar{v})b_1^2(t) - \frac{\bar{h}}{4}(|\lambda_1|\theta_2(t) + 2\bar{v})b_2^2(t). \tag{41}$$

In order to get an energy decrease, we chosse θ_1 such that the RHS of (41) is non-positive. In fact we choose θ_1 as follows:

$$\theta_1(t) \ge \frac{2\bar{v}}{\lambda_2}. \tag{42}$$

Note that this choice of θ_1 is always possible since $\frac{2\bar{v}}{\lambda_2} < 1$. Indeed $\frac{2\bar{v}}{\lambda_2} < 1$, because the subcritical flow condition (23) gives $\lambda_2 = \sqrt{g\bar{h}} + \bar{v} > 2\bar{v}$. Thus, we get the following result

Theorem 2. Let $t_k = kL/|\lambda_1|$, $k \in \mathbb{N}$. Assume that (28) holds, the initial condition $(\check{h}^0, \check{v}^0)$ is continuous in $]0, L[$, $(\check{v}_0, \check{v}_L)$ satisfies (40) and θ_1 satisfies (42). Then (24)-(26) has a unique solution $(\check{h}, \check{v})$ continuous in $[t_k, t_{k+1}] \times]0, L[$ satisfying the following energy estimate:

$$E(t_{k+1}) \leq (1 - \Theta^k) E(t_k), \tag{43}$$

where E is given by (30) and

$$\Theta^k = \min\left(\inf_{x \in]0,L[} \left(\frac{|\lambda_1|}{\lambda_2} \theta_2(t_k + \frac{L-x}{\lambda_2}) + \frac{2\bar{v}}{\lambda_2} \right), \inf_{x \in]0,L[} \left(\theta_1(t_k + \frac{x}{|\lambda_1|}) - \frac{2\bar{v}}{\lambda_2} \right) \right) \in [0,1[.$$

Proof: The existence and uniqueness of the solution follow by (27) and construction (33).

Integrating (41) from 0 to t_1, we have

$$E(L/|\lambda_1|) = E(0) - \frac{\bar{h}}{2} \int_0^{L/|\lambda_1|} (\lambda_2 \theta_1(t) - 2\bar{v}) b_1^2(t)\, dt - \frac{\bar{h}}{2} \int_0^{L/|\lambda_1|} (|\lambda_1|\theta_2(t) + 2\bar{v}) b_2^2(t)\, dt,$$

$$\leq E(0) - \frac{\bar{h}}{2} \int_0^{L/|\lambda_1|} (\lambda_2 \theta_1(t) - 2\bar{v}) \xi_1^2(0, |\lambda_1|t)\, dt$$

$$- \frac{\bar{h}}{2} \int_0^{L/\lambda_2} (|\lambda_1|\theta_2(t) + 2\bar{v}) \xi_2^2(0, L - \lambda_2 t)\, dt,$$

$$\leq E(0) - \frac{\bar{h}}{2|\lambda_1|} \int_0^L \left(\lambda_2 \theta_1(\frac{x}{|\lambda_1|}) - 2\bar{v} \right) \xi_1^2(0, x)\, dt$$

$$- \frac{\bar{h}}{2\lambda_2} \int_0^L \left(|\lambda_1|\theta_2(\frac{L-x}{\lambda_2}) + 2\bar{v} \right) \xi_2^2(0, x)\, dt,$$

$$\leq E(0) - \frac{\bar{h}}{2\lambda_2} \int_0^L \left(\lambda_2 \theta_1(\frac{x}{|\lambda_1|}) - 2\bar{v} \right) \xi_1^2(0, x)\, dt$$

$$- \frac{\bar{h}}{2\lambda_2} \int_0^L \left(|\lambda_1|\theta_2(\frac{L-x}{\lambda_2}) + 2\bar{v} \right) \xi_2^2(0, x)\, dt,$$

$$\leq E(0) - \frac{\bar{h}}{2} \int_0^L \left(\theta_1(\frac{x}{|\lambda_1|}) - \frac{2\bar{v}}{\lambda_2} \right) \xi_1^2(0, x)\, dt$$

$$- \frac{\bar{h}}{2} \int_0^L \left(\frac{|\lambda_1|}{\lambda_2} \theta_2(\frac{L-x}{\lambda_2}) + \frac{2\bar{v}}{\lambda_2} \right) \xi_2^2(0, x)\, dt,$$

$$\leq E(0) - \frac{\bar{h}}{2} \int_0^L \left[\xi_2^2(0, x) + \xi_1^2(0, x) \right] \Theta^0 dx, \tag{44}$$

where

$$\Theta^0 = \min\left(\inf_{x \in]0,L[} \left(\frac{|\lambda_1|}{\lambda_2} \theta_2(\frac{L-x}{\lambda_2}) + \frac{2\bar{v}}{\lambda_2} \right), \inf_{x \in]0,L[} \left(\theta_1(\frac{x}{|\lambda_1|}) - \frac{2\bar{v}}{\lambda_2} \right) \right).$$

We have $\Theta^0 \in [0,1[$, since we get $0 < \theta_1(\frac{x}{|\lambda_1|}) - \frac{2\bar{v}}{\lambda_2} < 1$ from (42) and the fact that $\frac{2\bar{v}}{\lambda_2} < 1$.

On the other hand, one has the following estimation

$$\xi_1^2(0,x) + \xi_2^2(0,x) = \left(\check{v}^0(x) - \check{h}^0(x)\sqrt{\frac{g}{\bar{h}}}\right)^2 + \left(\check{v}^0(x) + \check{h}^0(x)\sqrt{\frac{g}{\bar{h}}}\right)^2,$$

$$= 2(\check{v}^0(x))^2 + \frac{2g}{\bar{h}}(\check{h}^0(x))^2, = \frac{2}{\bar{h}}\left(\bar{h}(\check{v}^0(x))^2 + g(\check{h}^0(x))^2\right). \tag{45}$$

Therefore we deduce from (44)-(45) that

$$E(L/|\lambda_1|) \le E(0) - \Theta^0 \int_0^L \left(\bar{h}(\check{v}^0(x))^2 + g(\check{h}^0(x))^2\right) dx \le (1 - \Theta^0)E(0). \tag{46}$$

In order to generalize (46) with respect to time, we consider the time $t_k = kL/|\lambda_1|$ as initial condition. Then, we let

$$b_1(t) = \xi_1(t_k, |\lambda_1|(t - t_k)), \quad if \quad t \in]t_k, t_k + L/|\lambda_1|[,$$

$$b_2(t) = \xi_2(t_k, L - \lambda_2(t - t_k)), \quad if \quad t \in]t_k, t_k + L/\lambda_2[,$$

and

$$\Theta^k = \min\left(\inf_{x \in]0,L[}\left(\frac{|\lambda_1|}{\lambda_2}\theta_2(t_k + \frac{L-x}{\lambda_2}) + \frac{2\bar{v}}{\lambda_2}\right), \inf_{x \in]0,L[}\left(\theta_1(t_k + \frac{x}{|\lambda_1|}) - \frac{2\bar{v}}{\lambda_2}\right)\right) \in [0,1[.$$

And, by integrating from t_k to t_{k+1} and using the same arguments as for the interval $[0, t_1]$, the proof of Theorem 2 is finished. □

Remark 2.

1. *Using the weak formulation (31) and the fact that $C^0(]0, L[)$ is dense in $L^2(]0, L[)$, it is possible (using the arguments of (Goudiaby et al., -)) to prove that for initial data $(\check{h}^0, \check{v}^0)$ in $(L^2(]0, L[))^2$, the solution $(\check{h}, \check{v})$ of (24)-(26) satisfies (43) and the following regularity*

$$\left(\begin{matrix} \check{h} \\ \bar{h}\check{v} + \bar{v}\check{h} \end{matrix}\right), \quad \left(\begin{matrix} \check{v} \\ \bar{v}\check{v} + g\check{h} \end{matrix}\right) \in H(div, Q), \tag{47}$$

where $Q =]t_k, t_{k+1}[\times]0, L[$,

$$div \equiv \left(\frac{\partial}{\partial t}, \frac{\partial}{\partial x}\right) \quad and \quad H(div, Q) = \left\{V \in L^2(Q)^2; div\, V \in L^2(Q)\right\}.$$

2. *It is also possible to stabilize the reach by acting only on one free endpoint as in (Goudiaby et al., -).*
3. *Only the initial condition and the on-line measurements of the water levels at the endpoints are required to implement the feedback control law (40).*
4. *For an application need, in order to implement the controllers (40), one can use two underflow gates located at the left end ($x = 0$) and the right end ($x = L$) of the canal (see Fig 3). Denote by I_0 and I_L the gates opening. A relation between under flow gates opening and discharge is given as follows (see (De Halleux et al., 2003; Ndiaye & Bastin., 2004)).*

$$lh(t,0)v(t,0) = I_0(t)k_1\sqrt{2g(H_{up} - h(t,0))}, \tag{48}$$

$$lh(t,L)v(t,L) = I_L(t)k_2\sqrt{2g(h(t,L) - H_{down})}, \tag{49}$$

where, l is the width of the reach (m), k_1, k_2 are gate coefficients, $v(t,x) = \bar{v}(x) + \check{v}(t,x)$, $h(t,x) = \bar{h}(x) + \check{h}(t,x)$, H_{up} and H_{down} are the left and right water levels outside the canal, respectively. H_{up} and H_{down} are supposed to be constant and satisfy $H_{up} > h(t,0)$ and $h(t,L) > H_{down}$.

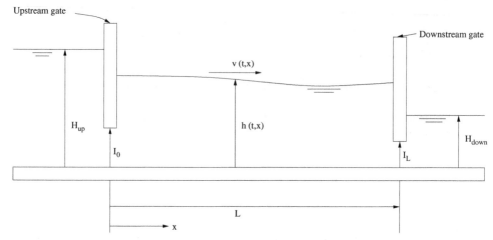

Fig. 3. A canal delimited by underflow gates

4. Building the controller for the cascade network

In this section, we use the idea of section 3.2, to build feedback control laws for the network.

4.1 Energy estimation and controllers building

Consider the energy of the network given by

$$E = \sum_{i=1}^{2} E_i, \quad E_i = \int_0^{L_i} \left(g\check{h}_i^2(t) + \bar{h}_i\check{v}_i^2(t) \right) dx. \tag{50}$$

Arguing as in section 3.1, from the weak formulation of (8), we deduce

$$
\begin{aligned}
\frac{1}{2}\frac{d}{dt}E(t) = &-\frac{\bar{h}_1\bar{v}_1}{2}\check{v}_{1,L_1}^2(t) - \frac{g\bar{v}_1}{2}\check{h}_1^2(t,L_1) - g\bar{h}_1\check{h}_1(t,L_1)\check{v}_{1,L_1}(t) \\
&+\frac{\bar{h}_1\bar{v}_1}{2}\check{v}_{1,0}^2(t) + \frac{g\bar{v}_1}{2}\check{h}_1^2(t,0) + g\bar{h}_1\check{h}_1(t,0)\check{v}_{1,0}(t) \\
&-\frac{\bar{h}_2\bar{v}_2}{2}\check{v}_{2,L_2}^2(t) - \frac{g\bar{v}_2}{2}\check{h}_2^2(t,L_2) - g\bar{h}_2\check{h}_2(t,L_2)\check{v}_{2,L_2}(t) \\
&+\frac{\bar{h}_2\bar{v}_2}{2}\check{v}_{2,0}^2(t) + \frac{g\bar{v}_2}{2}\check{h}_2^2(t,0) + g\bar{h}_2\check{h}_2(t,0)\check{v}_{2,0}(t).
\end{aligned} \tag{51}
$$

Using (13) and refering to figure 4, we express outgoing characteristic variables in terms of initial data and the solution at the endpoints and at the junction M at earlier times, i.e (14) is

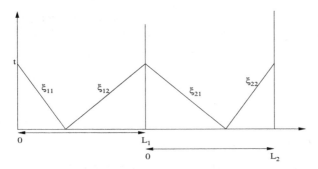

Fig. 4. Characteristic variables for the cascade network

satisfies with,

$$
b_1(t) = \begin{cases} \xi_{11}(0, |\lambda_{11}|t), & t \leq \frac{L_1}{|\lambda_{11}|}, \\ \xi_{11}(t - \frac{L_1}{|\lambda_{11}|}, L_1), & t \geq \frac{L_1}{|\lambda_{11}|}. \end{cases} \qquad
b_2(t) = \begin{cases} \xi_{22}(0, L_2 - \lambda_{22}t), & t \leq \frac{L_2}{\lambda_{22}}, \\ \xi_{22}(t - \frac{L_2}{\lambda_{22}}, 0), & t \geq \frac{L_2}{\lambda_{12}}. \end{cases}
$$

$$
b_3(t) = \begin{cases} \xi_{12}(0, L_1 - \lambda_{12}t), & t \leq \frac{L_1}{\lambda_{12}}, \\ \xi_{12}(t - \frac{L_1}{\lambda_{12}}, 0), & t \geq \frac{L_1}{\lambda_{12}}. \end{cases} \qquad
b_4(t) = \begin{cases} \xi_{21}(0, |\lambda_{21}|t), & t \leq \frac{L_2}{|\lambda_{21}|}, \\ \xi_{21}(t - \frac{L_2}{|\lambda_{21}|}, L_2), & t \geq \frac{L_2}{|\lambda_{21}|}. \end{cases}
$$

(52)

On the other hand, using (11) we also express the height at the boundaries and at the junction in terms of the flow velocity and outgoing characteristic variables:

$$
\check{h}_i(t, L_i) = \left(\xi_{i2}(t, L_i) - v_{i,L_i}(t) \right) \sqrt{\frac{\bar{h}_i}{g}},
$$

$$
\check{h}_i(t, 0) = \left(-\xi_{i1}(t, 0) + v_{i,0}(t) \right) \sqrt{\frac{\bar{h}_i}{g}}.
$$

(53)

Plugging (53) into (51), one gets

$$
\frac{1}{2}\frac{dE}{dt}(t) = a_1 \check{v}_{1,0}^2(t) - a_1 b_1(t)\check{v}_{1,0}(t) + c_1(t) + a_2 \check{v}_{2,L_2}^2(t) - a_2 b_2(t)\check{v}_{2,L_2}(t) + c_2(t)
$$

(54)

$$
+ a_3 \check{v}_{1,L_1}^2(t) - a_3 b_3(t)\check{v}_{1,L_1}(t) + c_3(t) + a_4 \check{v}_{2,0}^2(t) - a_4 b_4(t)\check{v}_{2,0}(t) + c_4(t)
$$

where

$$
a_1 = \bar{h}_1 \lambda_{12}, \qquad a_2 = \bar{h}_2 |\lambda_{21}|, \qquad a_3 = \bar{h}_1 |\lambda_{11}|, \qquad a_4 = \bar{h}_2 \lambda_{22},
$$

(55)

$$
c_1(t) = \frac{\bar{h}_1 \bar{v}_1}{2} b_1^2(t), \ c_2(t) = -\frac{\bar{h}_2 \bar{v}_2}{2} b_2^2(t), \ c_3(t) = -\frac{\bar{h}_1 \bar{v}_1}{2} b_3^2(t), \ c_4(t) = \frac{\bar{h}_2 \bar{v}_2}{2} b_4^2(t),
$$

$b_i, i = 1, 2, 3, 4$ are given by (52).

The flow conservation condition (8.c) is used to express $\breve{v}_{2,0}$ in terms of $\breve{v}_{1,L_1}$ and outgoing characteristic variables. From (8.c) and (53), one has

$$\breve{v}_{2,0}(t) = \alpha\breve{v}_{1,L_1}(t) + \beta b_3(t) + \delta b_4(t). \tag{56}$$

where

$$\alpha = \frac{|\lambda_{11}|}{\lambda_{22}}\sqrt{\frac{\bar{h}_1}{\bar{h}_2}}, \qquad \beta = \frac{\bar{v}_1}{\lambda_{22}}\sqrt{\frac{\bar{h}_1}{\bar{h}_2}}, \qquad \delta = \frac{\bar{v}_2}{\lambda_{22}}. \tag{57}$$

Thus, the last six terms of (54) can be expressed as follows:

$$a_3\breve{v}^2_{1,L_1}(t) - a_3 b_3(t)\breve{v}_{1,L_1}(t) + c_3(t) + a_4\breve{v}^2_{2,0}(t) - a_4 b_4(t)\breve{v}_{2,0}(t) + c_4(t)$$
$$= \sigma\breve{v}^2_{1,L_1}(t) + \gamma(t)\breve{v}_{1,L_1}(t) + \rho(t), \tag{58}$$

where

$$\sigma = (a_3 + \alpha^2 a_4) = \bar{h}_1|\lambda_{11}|\left(1 + \frac{|\lambda_{11}|}{\lambda_{22}}\right) \tag{59}$$

$$\gamma(t) = (2a_4\alpha\beta - \bar{h}_1|\lambda_{11}|)b_3(t) + \alpha(2a_4\delta - \bar{h}_2\lambda_{22})b_4(t)$$

$$= \bar{h}_1|\lambda_{11}|\left(\frac{2\bar{v}_1}{\lambda_{22}} - 1\right)b_3(t) + |\lambda_{11}|\sqrt{\bar{h}_1\bar{h}_2}\left(\frac{2\bar{v}_2}{\lambda_{22}} - 1\right)b_4(t) \tag{60}$$

$$\rho(t) = \left(a_4\delta^2 - \bar{h}_2\lambda_{22}\delta + \frac{\bar{h}_2\bar{v}_2}{2}\right)b_4^2(t) + \left(a_4\beta^2 - \frac{\bar{h}_1\bar{v}_1}{2}\right)b_3^2(t)$$
$$+ \beta(2a_4 - \bar{h}_2\lambda_{22})b_3(t)b_4(t)$$

$$= \frac{\bar{h}_2\bar{v}_2}{2}\left(\frac{2\bar{v}_2}{\lambda_{22}} - 1\right)b_4^2(t) + \frac{\bar{h}_1\bar{v}_1}{2}\left(\frac{2\bar{v}_1}{\lambda_{22}} - 1\right)b_3^2(t)$$

$$\tag{61}$$

$$+ \bar{v}_1\sqrt{\bar{h}_1\bar{h}_2}\left(\frac{2\bar{v}_1}{\lambda_{22}} - 1\right)b_3(t)b_4(t).$$

Taking into account (58), the energy law (54) becomes

$$\frac{1}{2}\frac{dE}{dt}(t) = a_1\breve{v}^2_{1,0}(t) - a_1 b_1(t)\breve{v}_{1,0}(t) + c_1(t) + a_2\breve{v}^2_{2,L_2}(t) - a_2 b_2(t)\breve{v}_{2,L_2}(t) + c_2(t)$$
$$+ \sigma\breve{v}^2_{1,L_1}(t) + \gamma(t)\breve{v}_{1,L_1}(t) + \rho(t). \tag{62}$$

If we prescribe the velocity at the boundaries as follows,

$$\breve{v}_{1,0}(t) = -\frac{b_1(t)}{2}\left(\sqrt{1 - \theta_1(t)} - 1\right),$$

$$\breve{v}_{2,L_2}(t) = -\frac{b_2(t)}{2}\left(\sqrt{1 - \theta_2(t)} - 1\right), \tag{63}$$

$$\breve{v}_{1,L_1}(t) = \frac{\gamma(t)}{2\sigma}\left(\sqrt{1 - \theta_3(t)} - 1\right),$$

where θ_1, θ_2, $\theta_3 : \mathbb{R}^+ \longrightarrow [0,1]$, it follows from Lemma 1 that

$$\frac{1}{2}\frac{dE}{dt}(t) = -\frac{b_1^2(t)}{4a_1}\theta_1(t) + c_1(t) - \frac{b_2^2(t)}{4a_2}\theta_2(t) + c_2(t) - \frac{\gamma^2(t)}{4\sigma}\theta_3(t) + \rho(t). \tag{64}$$

Let us calculate explicitely the RHS of (64). On the one hand, using (a_1, c_1) and (a_2, c_2) given in (55), we have

$$-\frac{b_1^2}{4a_1}\theta_1 + c_1 = -\frac{\bar{h}_1}{4}\left(\lambda_{12}\theta_1 - 2\bar{v}_1\right)b_1^2(t), \tag{65}$$

and

$$-\frac{b_2^2}{4a_2}\theta_2 + c_2 = -\frac{\bar{h}_2}{4}\left(|\lambda_{21}|\theta_2 + 2\bar{v}_2\right)b_2^2(t). \tag{66}$$

On the other hand, from (59)-(61) and using the fact that $\theta_3 \in \,]0,1]$ we have

$$-\frac{\gamma^2}{4\sigma}\theta_3 + \rho \leq \rho = \frac{\bar{h}_2\bar{v}_2}{2}\left(\frac{2\bar{v}_2}{\lambda_{22}} - 1\right)b_4^2(t) + \frac{\bar{h}_1\bar{v}_1}{2}\left(\frac{2\bar{v}_1}{\lambda_{22}} - 1\right)b_3^2(t)$$

$$+ \bar{v}_1\sqrt{\bar{h}_1\bar{h}_2}\left(\frac{2\bar{v}_1}{\lambda_{22}} - 1\right)b_3(t)b_4(t). \tag{67}$$

Since $\frac{2\bar{v}_2}{\lambda_{22}} < 1$, we get

$$\bar{v}_1\sqrt{\bar{h}_1\bar{h}_2}\left(\frac{2\bar{v}_2}{\lambda_{22}} - 1\right)b_3(t)b_4(t) \leq \frac{\bar{h}_1\bar{v}_1}{2}\left(1 - \frac{2\bar{v}_2}{\lambda_{22}}\right)b_3^2(t) + \frac{\bar{h}_2\bar{v}_1}{2}\left(1 - \frac{2\bar{v}_2}{\lambda_{22}}\right)b_4^2(t). \tag{68}$$

Combining (68) and (67), one has

$$-\frac{\gamma^2}{4\sigma}\theta_3 + \rho \leq \bar{h}_1\bar{v}_1\frac{(\bar{v}_1 - \bar{v}_2)}{\lambda_{22}}b_3^2(t) + \frac{\bar{h}_2}{2}\left(\frac{2\bar{v}_2}{\lambda_{22}}(\bar{v}_2 - \bar{v}_1) - (\bar{v}_2 - \bar{v}_1)\right)b_4^2(t). \tag{69}$$

Using (65), (66) and (69), the energy law (64) becomes

$$\frac{1}{2}\frac{dE}{dt}(t) \leq -\frac{\bar{h}_1}{4}\left((\lambda_{12}\theta_1 - 2\bar{v}_1)b_1^2(t) + 4\bar{v}_1\frac{(\bar{v}_2 - \bar{v}_1)}{\lambda_{22}}b_3^2(t)\right)$$

$$-\frac{\bar{h}_2}{4}\left(2(\bar{v}_2 - \bar{v}_1)\left(1 - \frac{2\bar{v}_2}{\lambda_{22}}\right)b_4^2(t) + (|\lambda_{21}|\theta_2 + 2\bar{v}_2)b_2^2(t)\right). \tag{70}$$

The way the steady state $(\bar{h}_1, \bar{v}_1, \bar{h}_2, \bar{v}_2)$ is chosen (see (6)), yields that

$$\bar{v}_2 \geq \bar{v}_1. \tag{71}$$

The function θ_1 satisfies a condition similar to (42), i.e

$$\theta_1(t) \geq \frac{2\bar{v}_1}{\lambda_{12}}. \tag{72}$$

Using the fact that $\frac{2\bar{v}_2}{\lambda_{22}} < 1$, (71) and (72), the RHS of (70) is non-positive. Thus we give the proof of Theorem 1.

4.2 Proof of theoreme 1

The existence and uniqueness of the solution follow by (11) and constructions (14).

Integrating (70) from 0 to t_1, one has

$$E(t_1) \leq E(0) - \frac{\bar{h}_1}{2} \int_0^{t_1} (\lambda_{12}\theta_1(t) - 2\bar{v}_1) b_1^2(t)dt - \frac{\bar{h}_1}{2} \int_0^{t_1} 4\bar{v}_1 \frac{(\bar{v}_2 - \bar{v}_1)}{\lambda_{22}} b_3^2(t)dt$$

$$- \frac{\bar{h}_2}{2} \int_0^{t_1} 2(\bar{v}_2 - \bar{v}_1)\left(1 - \frac{2\bar{v}_2}{\lambda_{22}}\right) b_4^2(t)dt - \frac{\bar{h}_2}{2} \int_0^{t_1} (|\lambda_{21}|\theta_2 + 2\bar{v}_2) b_2^2(t)dt.$$

$$\overset{(52)}{\leq} E(0) - \frac{\bar{h}_1}{2} \int_0^{\frac{L_1}{|\lambda_{11}|}} (\lambda_{12}\theta_1(t) - 2\bar{v}_1)\zeta_{11}^2(0, |\lambda_{11}|t)dt$$

$$- \frac{\bar{h}_1}{2} \int_0^{\frac{L_1}{\lambda_{12}}} 4\bar{v}_1 \frac{(\bar{v}_2 - \bar{v}_1)}{\lambda_{22}}\zeta_{12}^2(t, L_1 - \lambda_{12}t)dt$$

$$- \frac{\bar{h}_2}{2} \int_0^{\frac{L_1}{|\lambda_{21}|}} 2(\bar{v}_2 - \bar{v}_1)\left(1 - \frac{2\bar{v}_2}{\lambda_{22}}\right)\zeta_{21}^2(t, |\lambda_{21}|t)dt$$

$$- \frac{\bar{h}_2}{2} \int_0^{\frac{L_2}{\lambda_{22}}} (|\lambda_{21}|\theta_2(t) + 2\bar{v}_2)\zeta_{22}^2(0, L_2 - \lambda_{22}t)dt,$$

$$\leq E(0) - \frac{\bar{h}_1}{2} \int_0^{L_1} \left(\theta_1(\frac{x}{|\lambda_{11}|}) - \frac{2\bar{v}_1}{\lambda_{12}}\right)\zeta_{11}^2(0, x)dx$$

$$- \frac{\bar{h}_1}{2} \int_0^{L_1} 4\bar{v}_1 \frac{(\bar{v}_2 - \bar{v}_1)}{\lambda_{22}\lambda_{12}}\zeta_{12}^2(0, x)dx$$

$$- \frac{\bar{h}_2}{2} \int_0^{L_2} \left(\frac{|\lambda_{21}|}{\lambda_{22}}\theta_2(\frac{L_2 - x}{\lambda_{22}}) + \frac{2\bar{v}_2}{\lambda_{22}}\right)\zeta_{22}^2(0, x)dx$$

$$- \frac{\bar{h}_2}{2} \int_0^{L_2} 2\frac{(\bar{v}_2 - \bar{v}_1)}{\lambda_{22}}\left(1 - \frac{2\bar{v}_2}{\lambda_{22}}\right)\zeta_{21}^2(0, x)dx.$$

$$\leq E(0) - \frac{\bar{h}_1}{2} \int_0^{L_1} \left[\zeta_{11}^2(0, x) + \zeta_{12}^2(0, x)\right]\Gamma_1^0 dx$$

$$- \frac{\bar{h}_2}{2} \int_0^{L_2} \left[\zeta_{22}^2(0, x) + \zeta_{21}^2(0, x)\right]\Gamma_2^0 dx, \tag{73}$$

where

$$\Gamma_1^0 = \min\left(\inf_{x \in]0,L_1[}\left(\theta_1(\frac{x}{|\lambda_{11}|}) - \frac{2\bar{v}_1}{\lambda_{12}}\right), 4\bar{v}_1 \frac{(\bar{v}_2 - \bar{v}_1)}{\lambda_{22}\lambda_{12}}\right),$$

$$\Gamma_2^0 = \min\left(\inf_{x \in]0,L_2[}\left(\frac{|\lambda_{21}|}{\lambda_{22}}\theta_2(\frac{L_2 - x}{\lambda_{22}}) + \frac{2\bar{v}_2}{\lambda_{22}}\right), 2\frac{(\bar{v}_2 - \bar{v}_1)}{\lambda_{22}}\left(1 - \frac{2\bar{v}_2}{\lambda_{22}}\right)\right).$$

Arguing as for (45), we get

$$\xi_{i1}^2(0,x) + \xi_{i2}^2(0,x) = \frac{2}{\bar{h}_i}\left(\bar{h}_i(\check{v}_i^0(x))^2 + g(\check{h}_i^0(x))^2\right). \tag{74}$$

Therefore, using (74) in (73), one has

$$E(t_1) \le (1 - \Theta^0)E(0) \tag{75}$$

where

$$\Theta^0 = \min\left(\Gamma_1^0, \Gamma_2^0\right) \in [0,1[, \quad \text{since} \quad 0 < \theta_1(\frac{x}{|\lambda_{11}|}) - \frac{2\bar{v}_1}{\lambda_{12}} < 1.$$

In order to generalize (75) with respect to time, we consider the time $t_k = kT$ as initial condition, with T given by (17). Then, we let

$$b_1(t) = \xi_{11}(t_k, |\lambda_{11}|(t - t_k)), \quad t \in]t_k, t_k + L_1/|\lambda_{11}|[,$$

$$b_2(t) = \xi_{22}(t_k, L_2 - \lambda_{22}(t - t_k)), \quad t \in]t_k, t_k + L_2/\lambda_{22}[,$$

$$b_3(t) = \xi_{12}(t_k, L_1 - \lambda_{12}(t - t_k)), \quad t \in]t_k, t_k + L_1/\lambda_{12}[,$$

$$b_3(t) = \xi_{21}(t_k, |\lambda_{21}|(t - t_k)), \quad t \in]t_k, t_k + L_2/|\lambda_{21}|[,$$

$$\Gamma_1^k = \min\left(\inf_{x \in]0,L_1[}\left(\theta_1(t_k + \frac{x}{|\lambda_{11}|}) - \frac{2\bar{v}_1}{\lambda_{12}}\right), 4\bar{v}_1\frac{(\bar{v}_2 - \bar{v}_1)}{\lambda_{22}\lambda_{12}}\right),$$

$$\Gamma_2^k(x) = \min\left(\inf_{x \in]0,L_2[}\left(\frac{|\lambda_{21}|}{\lambda_{22}}\theta_2(t_k + \frac{L_2 - x}{\lambda_{22}}) + \frac{2\bar{v}_2}{\lambda_{22}}\right), 2\frac{(\bar{v}_2 - \bar{v}_1)}{\lambda_{22}}\left(1 - \frac{2\bar{v}_2}{\lambda_{22}}\right)\right).$$

and

$$\Theta^k = \min\left(\Gamma_1^k, \Gamma_2^k\right) \in [0,1[.$$

Therefore, by integrating from t_k to t_{k+1} and using the same arguments as for the interval $[0, t_1]$, the proof of Theorem 1 is completed. □

5. Numerical results

Numerical results are obtained by using a high order finite volume method (see Leveque. (2002); Toro. (1999)).

5.1 A numerical example for a single reach

In this section, we illustrate the control design method on a canal with the following parameters. Lenght $L = 500m$, width $l = 1m$. The steady state is $\bar{q}(x) = 1m^3s^{-1}$ and $\bar{h}(x) = 1m$ and the initial condition is $h(0,x) = 2m$ and $q(0,x) = 3m^3s^{-1}$. The spatial step size is $\Delta x = 10m$ and the time step is $\Delta t = 1s$. We also set $H_{up} = 2.2m$ and $H_{down} = 0.5m$ and use relations (48)-(49) for gates opening.

We have tested a big perturbation in order to investigate the robusness and the flexibility of the control method. One sees that the bigger the θ's are, the faster the exponential deacrease is (Fig 5). Increasing θ's also produces some oscillations of the gates opening with heigh frequencies (Fig 6). We then notice that for the gates opening, choosing θ's between 0.5 and 0.7 gives a quite good behaviour of the gates opening (Fig 7-(b)). Generally, depending on the control action (gates, pumps etc) used, we can have a wide possibilities of choosing the θ's.

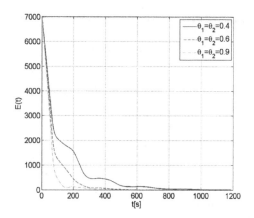

Fig. 5. Energy evolution for different values of θ_1 and θ_2.

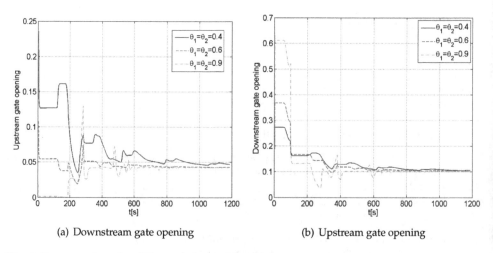

(a) Downstream gate opening   (b) Upstream gate opening

Fig. 6. Gate openings for different values of θ_1 and θ_2.

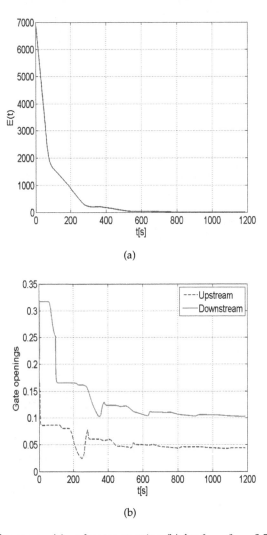

Fig. 7. Evolution of the energy (a) and gates opening (b) for $\theta_1 = \theta_2 = 0.5$.

5.2 A numerical example for the cascade network

We consider two reaches of Lenght $L_1 = L_2 = 1000m$, width $l = 1m$. The steady state is $\bar{q}_1(x) = 1.5m^3s^{-1}$, $\bar{h}_1(x) = 1.5m$ and $\bar{h}_2(x) = 1m$ and the initial condition is $h_1(0, x) = 2m$, $q_1(0, x) = 3m^3s^{-1}$, $h_2(0, x) = 1.5m$, $q_2(0, x) = 3m^3s^{-1}$. The spatial step size is $\Delta x = 10m$ and the time step is $\Delta t = 1s$. We also set $H_{up} = 3m$ and $H_{down} = 0.5m$. We have noticed as in the case of one single reach, that the bigger the θs are, the faster the exponential decrease is. In figure (8), we have plotted the energy decay and the gates opening for $\theta_1 = \theta_2 = \theta_3 = 0.7$. Although, the perturbations for reach 1 and 2 are different, the controllers act in such a way to drive the perturbations to zero simultaneously (Fig (9)).

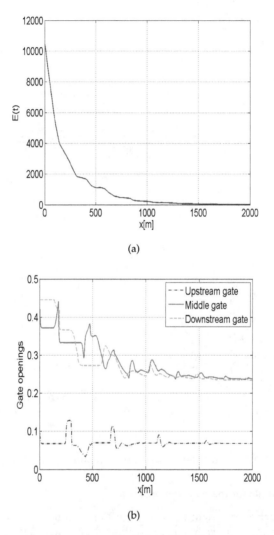

(a)

(b)

Fig. 8. Energy evolution (a) and gates opening (b) for $\theta_1 = \theta_2 = \theta_3 = 0.7$.

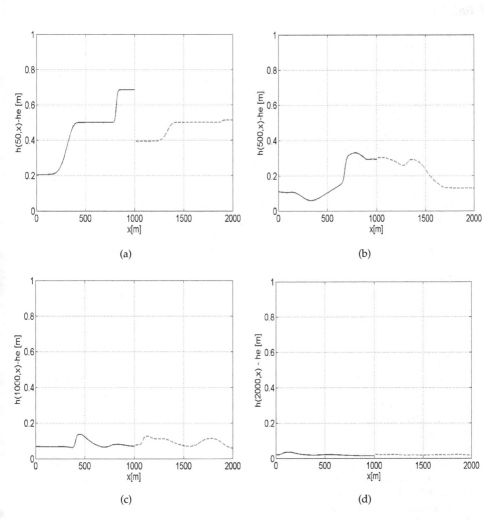

Fig. 9. Deviation of water height at instants $t = 50$ (a), $t = 500$ (b), $t = 1000$ (c) and $t = 2000$ (d), for $\theta_1 = \theta_2 = \theta_3 = 0.7$.

6. Acknowledgements

This work is supported by International Science Programme (ISP, Sweden) and AIRES Sud projet EPIMAT.

7. References

Balogun, O.; Hubbard, M. & De Vries, J. J. (1988). Automatic control of canal flow using linear quadratic regulator theory, *Journal of hydraulic engineering*, Vol. 114, No. 1, (1988), 75-102.

Bastin, G.; Bayen, A. M.; D'apice, C.; Litrico, X. & Piccoli, B. (2009). Open problems and research perspectives for irrigation channels, *Networks and Heterogenous Media*, Vol. 4, No. 2, (2009), 1-4.

Bastin, G.; Coron, J. M.; D'Andréa-Novel, B. (2009). On lyapunov stability of linearised Saint-Venant equations for a sloping channel, *Networks and Heterogenous Media*, Vol. 4, No. 2, (2009), 177-187.

Cen, L. H.; Xi, Y. G. (2009). Stability of Boundary Feedback Control Based on Weighted Lyapunov Function in Networks of Open Channels, *Acta Automatica Sinica*, Vol. 35, Issue. 1, (2009), 97-102.

Chen, M.; Georges, D. & Lefevre, L. (2002).Infinite dimensional LQ control of an open-channel hydraulic system, ASCC, Singapoure (2002).

De Halleux, J.; Prieur, C.; Coron, J. M.; D'Andréa-Novel, B. & Bastin, G. (2003) Boundary feedback control in networks of open channels, *Automatica*, Vol. 39, Issue. 8, (2003), 1365-1376.

De Saint-Venant, B. (1871)Théorie du mouvement non-permanent des eaux avec application aux crues des rivières et à l'introduction des marées dans leur lit, *Comptes Rendus Academie des Sciences*, Vol. 73, (1871), 148-154, 237-240.

Goudiaby, M. S.; Sene, A. & Kreiss, G. An algebraic approach for controlling star-like network of open canals, *preprint*.

Gowing, J. (1999), Limitations of water-control technology, *Agriculture Water Management*, Vol. 79, (1999), 95-99.

Leugering, G. & Georg Schmidt, E. J. P. (2002) On the modelling and stabilization of flows in networks of open canal, *SIAM J. Control Optim.*, Vol. 41, No 1, (2002), 164-180.

Leveque, R. J. (2002), *Finite Volume Methods for Hyperbolic Problems*, Cambridge Texts in Applied Mathematics, USA (2002).

T. Li, (2005) Exact boundary controlability of unsteady flows in a network of open canals, *Math. Nachr.*, Vol. 278, No 3, (2005), 278-289.

Litrico, X.; Fromion, V.; Baume, J. P. & Rijo, M. (2003) Modeling and PI control of an irrigation canal, *European Control Conference ECC* (2003), 6p.

Malaterre, P.O. (1998), PILOTE Linear quadratic optimal controller for irrigation canals, *Journal of irrigation and drainage engineering*, Vol. 124, No 4, (1998), 187-193.

Ndiaye, M. & Bastin, G. (2004), Commande fontiere adptative d'un bief de canal avec prélévements inconnus, *RS-JESA*, Vol. 38, (2004), 347-371.

E. F. Toro, *Riemann Solvers and Numerical Methods for Fluid Dynamics: A Practical Introduction* Springer-Verlag, Berlin (1999).

Weyer, E. (2003), LQ control of irrigation channel, *Decision and Control*, Vol. 1, (2003), 750-755.

Xu, C. Z. & Sallet, G. (1999) Proportional and integral regulation of irrigation canal systems governed by the Saint-Venant equation, *Proceedings of the 14th world congress IFAC*, pp. 147-152, Beijing, July 1999, Elsevier.

8

Performance of Smallholder Irrigation Schemes in the Vhembe District of South Africa

Wim Van Averbeke
Tshwane University of Technology
South Africa

1. Introduction

South Africa needs to raise employment and reduce poverty, particularly among rural African people. The New Growth Path released by the government in November 2010 was a response to the persistent unemployment problem. It aims to create five million new jobs by 2020. The New Growth Path intends to create 300 000 of these new jobs through the establishment of smallholder farmer schemes (Department of Economic Development, 2010). This suggests that policymakers believe that smallholder scheme development can create a substantial number of new employment opportunities in South Africa. However, the performance of the smallholder schemes that have been set up as part of the post-democratisation land reform programme has been dismal (Umhlaba, 2010). Assessments of smallholder irrigation schemes indicated that many of them also performed poorly (Bembridge, 2000; Machete et al., 2004; Tlou et al., 2006; Mnkeni et al., 2010). Yet, in water-stressed South Africa, expanding smallholder irrigation is one of the obvious options to trigger rural economic development. Elsewhere in the world, particularly in Asia, investment in irrigation was a key ingredient of the green revolution, which lifted large numbers of rural Asians out of poverty and created conditions that were conducive for the industrial and economic development that has occurred (Turral et al., 2010). A similar development trajectory has been recommended for South Africa and other parts of Sub-Saharan Africa (Lipton, 1996). So far, the developmental impact of smallholder irrigation in Sub-Saharan Africa has been limited (Inocencio et al., 2007).

In this chapter, associations between selected performance indicators and attributes of smallholder irrigation schemes in the Vhembe District of Limpopo Province, South Africa, are examined. For the purpose of this chapter, smallholder irrigation scheme was defined as an as an agricultural project that was constructed specifically for occupation by African farmers and that involved multiple holdings, which depended on a shared distribution system for access to irrigation water and in some cases also on a shared water storage or diversion facility. In 2011, there were 302 smallholder irrigation schemes in South Africa with a combined command area of 47 667 ha.

The objective of this chapter was to identify factors that had a significant effect on smallholder irrigation scheme performance. Knowledge of such factors could assist effective location and design of new schemes. Before focussing on the study itself, it was deemed important to provide a background to African smallholder agriculture in South Africa in general and smallholder irrigation scheme development in particular. This is done in the first part of this chapter. The study area and the materials and methods used to assess the population of smallholder irrigation schemes in the study area are presented in the second part. The results of the study are presented next. First, the characteristics of population of smallholder schemes in the study area are described. Next, the associations between a selection of smallholder scheme characteristics and four performance indicators are examined statistically. In the last part the results of the study are discussed and interpreted.

2. African smallholder agriculture and irrigation in South Africa: A brief history

2.1 African smallholder agriculture

In South Africa, traditional agriculture was disturbed in major ways by military and political subjugation of the different African tribes during the nineteenth century, followed by land dispossession, segregation and separation. These processes restricted the area where Africans held farm land to relatively small parts of the country, which combined covered about 13% of the total land area in 1994 (Vink & Kirsten, 2003). Over time, the territories in which Africans held land have been referred to as Native Areas, Bantu Areas, Bantustans and homelands but in this text the term homelands will be used to reflect the situation just prior to democratisation of South Africa in 1994. From when they were created in 1913, the homelands have been characterised by high rural population densities, small individual allotments of arable land and shared access to rangeland. The rangeland that was available to communities was inadequate to support sufficient livestock to meet even the most basic requirements of African homesteads in terms of draught power, milk, wool, meat and social needs (Lewis, 1984; Bundy, 1988; Mills & Wilson, 1952). African homesteads diversified their livelihoods in response to the lack of room to reproduce their land-based lifestyles. Until about 1970, migrant remittances, mostly from male members who worked in mines and cities, supported the reproduction of African rural homesteads (Beinart, 2001). From 1970 to 1990, income earned from employment inside the homelands became important (Leibbrandt & Sperber, 1997; Beinart, 2001). Homelands received substantial budget allocations from separatist South Africa to attend to local economic and social development. Employment was created in education, bureaucracy and business (Beinart, 2001). From 1990 onwards, rural homesteads increasingly depended of claiming against the state, in the form of old-age pensions and child support grants (Shah et al., 2002; Van Averbeke & Hebinck, 2007; Aliber & Hart, 2009; De Wet, 2011).

Despite the lack of room to farm, agricultural activities remained central in the livelihood strategy of a majority of rural homesteads until about 1950, even though the proportional and nominal contribution of agriculture to homestead income had been in decline for much longer (Houghton, 1955; Tomlinson Commission, 1955; Bundy, 1988). After 1950, rural African homesteads progressively withdrew from cultivating their arable allotments. For example, in Ciskei the cultivation of arable land dropped from an average of 82% in 1950 (Houghton, 1955) to 10% in 2006 (De Wet, 2010). Whilst the decline in cultivation has not necessarily been as dramatic in all homeland areas as it has been in Ciskei, the trend has been universal. As a result, for most African rural homesteads farming has become a livelihood activity that is of secondary importance. In 2006, of the 1.3 million African households with access to an arable allotment, only 8% listed agriculture as their main source of food. For less than one in ten households (9%) agriculture was a source of monetary income, with 3% referring to farming as their main source of income and 6% as an additional source of income. For the large majority, farming was merely an additional source of food (Aliber & Hart, 2009; Vink & Van Rooyen, 2009).

The deterioration of African farming received government attention from 1917 onwards (Beinart, 2003). Initial interventions were focused on the conservation of the natural resource base in the homelands. Land use planning, conservation of arable land using erosion control measures, rotational grazing using fenced grazing camps and livestock reduction schemes were some of the important measures taken by the state to check natural resource degradation (Beinart, 2003). More land was made available to reduce growing landlessness but these interventions had no positive impact on agricultural production. Overcrowding as the principal cause for the inability of African smallholders to produce enough to feed themselves, let alone make a living of the land, was pointed out by the Commission for the Socio-Economic Development of the Bantu Areas within the Union of South Africa (1955). This Commission was referred to as the Tomlinson Commission (1955) after Professor F.R. Tomlinson, who was its Chairperson. The Tomlinson Commission (1955) proposed the partial depopulation of the homelands to avail enough room for those who were to remain on the land to make a living from full-time farming. Expansion of irrigated farming, by upgrading existing irrigation schemes and establishing new schemes in the homelands, was one of the strategies proposed by the Tomlinson Commission (1955) to create new opportunities for African homesteads to make a full-time living from smallholder agriculture. The Tomlinson Commission (1955) had identified that smallholders on some of the existing irrigation schemes were making a decent living on irrigated plots of about 1.28 ha combined with access to enough grazing land to keep a herd of six cattle, which was the minimum number required for animal draught power. The master plan of the Tomlinson Commission (1955) to reduce the population in the homelands and establish economically viable farm units for African homesteads was never implemented. The proposal to expand smallholder irrigation did receive attention. It played an important role in irrigation scheme development in the Vhembe District, which will be discussed later.

2.2 African smallholder irrigation schemes

The use of irrigation by African farmers in South Africa appears to have two centres of origin. One of these centres was the Ciskei region of the Eastern Cape, where technology transfer from colonialists to the local people, resulted in the adoption of irrigated agriculture by African peasants (Bundy, 1988). These early smallholder irrigation developments were mostly private or mission station initiatives and involved river diversion. Most of these early African irrigation initiatives in the Eastern Cape did not last long (Houghton, 1955, Bundy, 1988). The other centre of origin was located in what is now the Vhembe District. Evidence of African irrigation in this area was provided by Stayt (1968), who conducted anthropological research among the Venda during the late nineteen-twenties and published the first account of his work in 1931. Box 1 cites Stayt's reference to African irrigation in Vhembe. This reference to early African irrigation in Vhembe contained in Box 1 is significant for two reasons, namely the apparent use of irrigated agriculture by local African people before exposure to European colonialists and their continued use, or at least re-adoption, of irrigated agriculture using stream diversion during the nineteen-thirties,. This suggests local interest, knowledge and affinity for the use of irrigation as a way of intensifying crop production.

In the northwest of Vendaland there are traces of some very ancient occupation. Colonel Piet Moller, who was an early settler in the Zoutpansberg, has found what he considers indisputable evidence of ancient irrigation works. Most of the old furrows are near Chepisse and it appears that the water was diverted from a small stream there in a series of furrows to a distance of about four and a half miles south. Traces of furrows are also discernable at Sulphur Springs, and at several places by the Nzhelele river, where some of them have been reopened and are utilised by the BaVenda to-day. Colonel Moller says that when he first came across these some forty years ago (around 1880), there was no doubt about their antiquity; to-day they are very difficult to trace, as roads, modern agriculture, and furrows have altered the face of the country considerably and have particularly hidden the ancient workings.

Box 1. Reference to African irrigation in Vhembe (Stayt, 1968)

The Tomlinson Commission (1955) also identified the northern parts of South Africa as the area where smallholder irrigation schemes were functioning best, as is evident from its statement reproduced in Box 2.

Among the various systems and types of settlement in the Bantu Areas, irrigation farming is undoubtedly the only form of undertaking in which, under European leadership and control, the Bantu have shown themselves capable of making a full-time living from farming, and of making advantageous use of the soil for food production.

The interest shown by Bantu in irrigation farming varies from one locality to another. In some parts of the Transvaal (here reference is made to areas that are now part of Limpopo Province), the Bantu are so enthusiastic that they offer their labour free to construct canals to lead water from streams for the irrigation of their land, while in the Transkei and Ciskei (now part of the Eastern Cape), on the contrary, interest has waned to such a degree, that existing schemes have fallen into disuse.

Box 2. Reference to the performance of African smallholders on irrigation schemes during the period 1950-52 by the Tomlinson Commission (1955)

In 1952, when the Tomlinson Commission completed its data collection, it identified 122 smallholder irrigation schemes covering a total of 11 406 ha. This irrigated area was held by 7 538 plot holders, each holding a plot with an average size of 1.513 ha. All of these were river diversion schemes but it is not clear whether the water conveyance and distribution systems were lined or not. The Tomlinson Commission (1955) did distinguish between what appeared to be indigenous and state controlled irrigation projects, identifying state controlled schemes as performing considerably better than those controlled by African farmers themselves (Box 3).

The 'European control' mentioned in Box 3 referred to a set of institutional arrangements imposed by the state, which regulated allocation of water to farmers and land use, including choice of crops, and the provision of technical advice and marketing assistance for the crops that were prescribed to farmers. In line with this observation, the Tomlinson Commission (1955) recommended the construction of new smallholder irrigation schemes and the upgrading of existing schemes as a smallholder development strategy. The Tomlinson Commission (1955) identified a total area of 54 051 ha that had the potential for irrigation development in Bantu Areas and estimated that exploitation of this potential could enable the settlement of 36 000 farmer families, representing approximately 216 000 people. The Tomlinson Commission (1955) recommended that irrigation scheme development should occur in the form of simple canal schemes using river diversion by means of a weir and that uniform regulations should be applied to the running of these schemes. One of these regulations was that ownership and control over tribal land identified for irrigation scheme development needed to be transferred to the state before construction of the scheme. Another was that homesteads would be allocated plots that were 1.28 ha to 1.71 ha in size, as these were deemed adequate to provide for a livelihood based on full-time farming. A third was the enforcement of specified production systems on smallholder irrigation schemes. These production systems were to be designed, enforced and supported by state-appointed superintendents. Farmers who settled on these schemes held their plots under Trust tenure. This form of tenure provided the state with the necessary powers to prescribe land use and to expel and replace farmers whose practices did not comply with these prescriptions. In selected cases the state effectively used these powers to enforce the overall objectives of the schemes by evicting poorly performing families (Van Averbeke, 2008). This authoritarian and paternalistic approach by the state was not limited to irrigation schemes settled by Africans. The same approach had been used on state schemes established for settlement by white farmers during the Great Depression and WWII period (Backeberg and Groenewald, 1995).

The Commission collected details of the production achieved on the controlled Olifants River irrigation scheme and the uncontrolled Njelele River scheme (Vhembe District). The average size holding were 1.53 morgen (1.3 ha) and 1.71 morgen (1.5 ha), respectively, and other physical factors were approximately equal. It was found that the average income per settler on the Olifants scheme was £110.69 as compared with £28.79 on the Njelele scheme. The average yield of grain of all sorts was 47.07 bags (4270 kg) (fil in) per settler on the Olifants, as against 9.2 bags (835 kg) on the Njelele scheme. This is a clear indication that irrigation schemes for Bantu are successful when under efficient control and guidance and that the average Bantu family on 1.5 morgen (1.28 ha) under such schemes, can make a gross income of £110.7 per annum, which renders it unnecessary for members of the family to seek employment elsewhere to supplement the family income.

Box 3. Comparison of the performance of African smallholders on indigenous irrigation schemes with those on irrigation schemes under state control (Tomlinson Commission, 1955)

Construction of smallholder canal schemes in South Africa was continued until the nineteen-seventies. The 2011 update of the smallholder irrigation scheme data base created by Denison and Manona (2007) indicated that there were 74 smallholder canal schemes left in South Africa. Sixty-seven of these were operational, six were not operational and of one scheme the operational status was not known. The combined command area of existing gravity-fed canal schemes was 11 966.2 ha, which represented 25.1% of the total smallholder irrigation scheme command area in South Africa. Surface irrigation was also practised on 20 schemes that used pumping, sometimes in combination with gravity. Among these 20 schemes, 14 of were operational and six were not. Combined they had a command area of 4 113.7 ha, 8.6% of the total.

From the nineteen-seventies onwards, the design of smallholder irrigation schemes in South Africa was influenced by the modernisation paradigm. This paradigm was based on the belief that modern, capital-intensive infrastructure, to be paid for by the intensive production of high-value crops, could lift smallholders out of poverty (Faurès *et al.*, 2007). Pumping and overhead irrigation became the norm in smallholder irrigation scheme development in South Africa. In 2011, there were 175 smallholder irrigation schemes that used overhead irrigation. Combined they had a command area of 27 757.6 ha, 58.2% of the total. Among these 175 schemes, 111 were operational, 59 were not and of five the operational status was not known. Pumped overhead schemes covered a total command area of 16 497.1 ha, gravity-fed overhead schemes 4 451 ha and schemes where gravity and pumping occurred in combination had a total command area of 6 903.5 ha.

Distinctive of the modernisation paradigm in smallholder irrigation scheme development was the establishment of large projects. In many of the large smallholder schemes that were constructed in South Africa, the design was characterised by functional diversification and centralisation of scheme management. Typically, these large schemes were designed to perform three functions, namely a commercial function, a commercial smallholder development function and a subsistence function. The commercial function was performed by allocating a substantial part of the scheme area to a central unit that was farmed as an estate. Farming on this estate used management and labour (Van Averbeke *et al.*, 1998). The commercial smallholder development function was implemented by allocating a limited number of 'mini-farms' to selected African homesteads, who were judged to have the aptitude to make a success of small-scale commercial agriculture. These mini-farms ranged between 5 ha and 12 ha in size. (Van Averbeke *et al.*, 1998), The subsistence function was put into practice by providing large numbers of African homesteads with access to food plots, ranging from 0.1 ha to 0.3 ha in size (Van Averbeke *et al.*, 1998). In some instances complex arrangements had to be made to implement this multi-functional design, because land holders had to be compensated for handing over their dryland allotments to create room for the central unit estate. A good example was the 2 830 ha Ncora Irrigation Scheme, established in 1976 in the Transkei region of the Eastern Cape. In return for availing their allotments to the scheme, the 1 200 existing land holders at Ncora were offered the right to 0.9 ha of irrigation land. They were given the choice of farming the entire allocation themselves or handing over two-thirds of their allotment to the central unit and remain with a 0.3 ha plot for own use. The latter option provided land holders with production inputs free of charge and an annual dividend derived from the profits made by the central unit. Management of these large schemes was centralised and in the hands of specialised parastatals established by homeland governments (Van Rooyen & Nene, 1996; Van Averbeke et al., 1998; Lahiff, 2000). The financial viability of this type of smallholder schemes was dependent on the performance of the central unit. Records show that the financial performance of these central units never met the predictions (Van Averbeke *et al.*, 1998). State subsidies were persistently required to keep these schemes afloat. Taking an extreme example, in 1995, the central unit of Ncora Irrigation Scheme required a budget of R21.3 million. It had 650 employees at a cost R16.6 million and operational costs amounting to R4.8 million. The income of the central unit in 1995 was R2.8 million, way short of even meeting its operational costs.

Following the democratisation of South Africa in 1994, the provincial governments decided to dismantle the agricultural homeland parastatals and transfer the management of smallholder irrigation schemes to the farmer communities who benefitted from them. Elsewhere in the world, a similar process, referred to as 'Irrigation Management Transfer' (IMT) had been occurring. Reducing public expenditure on irrigation, improving productivity of irrigation and stabilising of deteriorating irrigation systems were the three main reasons why IMT was implemented by governments (Vermillion, 1997). In South Africa, the dismantling of homeland parastatals and IMT proceeded very swiftly. It started in 1996 in the Eastern Cape and ended in 1998 in Limpopo Province. IMT affected all projects where parastatals were offering services to smallholders. Its effects were most strongly felt on the large, modern smallholder irrigation schemes, because these projects were the most complex to manage. Having been centrally managed from inception, levels of dependency on external management among farmers on these schemes were exceptionally high (Van Averbeke et al., 1998). Farming collapsed as soon as IMT had been implemented on these schemes (Bembridge, 2000; Laker, 2004). Small irrigation schemes, particularly the canal schemes, were more resilient and continued to operate, albeit at reduced levels (Kamara et al., 2001; Machete et al., 2004).

Besides IMT, the nineteen-nineties also saw the establishment of several new smallholder irrigation schemes. Conceptually, these new schemes were aligned with the Reconstruction and Development Programme (RDP). This Programme was the national development framework that applied at that time. It was aimed at eradicating poverty and improving the quality of life among poor African people in rural areas and informal urban settlements. Irrigation development focused on improving food security at community or group level and favoured the establishment of small schemes. In 2006, Denison and Manona (2007) identified 62 smallholder irrigation schemes that were established during this era, but combined they only covered 2 383 ha, clearly indicating their limited size (38.4 ha on average). Typically, these projects used mechanical pump and sprinkler technology to extract and apply irrigation water.

When GEAR (Growth, Employment and Redistribution) superseded the RDP as the overall development policy of South Africa, the strategy to eradicate poverty shifted from funding community-based projects to pursuing economic growth through private sector development. Existing irrigation schemes were identified as important resources for the economic development of the rural areas, but they required revitalisation first. The Revitalisation of Smallholder Irrigation Schemes (RESIS) of the Limpopo Province stood out for its comprehensiveness. The RESIS programme evolved from the WaterCare programme, which was launched in 1998 and ran for five years (Denison and Manona, 2007). The WaterCare programme was aimed at revitalizing selected smallholder irrigation schemes in the Province, not only infrastructurally, but also in terms of leadership, management and productivity. Using a participatory approach, WaterCare involved smallholder communities in planning and decision making and provided training to enable these communities to take full management responsibility over their schemes (Denison & Manona, 2007). In February 2000, Mozambique and the Limpopo Province were ravaged by cyclone Conny (Christie & Hanlon, 2001). Heavy rains caused widespread floods and damage to roads, bridges and also to the weirs that provided water to many of the smallholder canal schemes (Khandlhela & May, 2006). Declared a disaster area, the Limpopo Province was allocated special funding to repair the damage to its infrastructure, providing impetus to the WaterCare programme. In 2002, the Limpopo Province broadened the scope of its irrigation scheme rehabilitation intervention by launching a comprehensive revitalisation programme, called RESIS (REvitalisation of Smallholder Irrigation Schemes). RESIS adopted the participatory approach of the WaterCare programme, but planned to revitalise all smallholder schemes in the Province (Denison & Manona, 2007). As was the case in the WaterCare programme, RESIS combined the reconstruction of smallholder irrigation infrastructure with the provision of support to enable effective IMT. In support of IMT, the programme dedicated one-third of the revitalisation budget to capacity building among farmers. Guidelines for the sustainable revitalisation of smallholder irrigation schemes, which covered the building of capacity among irrigator communities were developed by De Lange et al.(2000). RESIS also sought to enhance commercialisation of the smallholder farming systems on the schemes, in order to improve the livelihood of plot holder homesteads (Van Averbeke, 2008).

During the WaterCare programme and the first phase of RESIS (1998-2005), the emphasis was primarily on the rehabilitation of the existing scheme infrastructure and on sustainable IMT, and less on commercialisation. Canal schemes that were revitalised during this phase remained canal schemes. However, in 2005, commercialisation became the principal development objective of RESIS. The shift in emphasis was probably influenced by the Black Economic Empowerment (BEE) strategy that was introduced in South Africa, first in the mining sector and later on also in other sectors of the economy, including agriculture (Van Averbeke, 2008). Nationally, the BEE strategy was aimed at increasing the share of black people in the economy and it emphasized entrepreneurship. In 2005, the Limpopo Department of Agriculture launched the second phase of RESIS, named RESIS-RECHARGE. The Department equated canal irrigation with subsistence farming and inefficient water use. Consequently, it discouraged and later on rejected revitalisation of canal infrastructure. Instead it funded the transformation of canal schemes into schemes that used modern irrigation technology, such as micro-irrigation, centre pivot and floppy sprinkler systems. Implementation of these new irrigation systems obliterated existing plot boundaries. To get production on these revitalised modern schemes on a commercial footing, the Department engaged the services of a strategic partner in the form of a commercial farmer, who was tasked with running the entire operation. Plot holders were compensated for availing their land holdings by means of dividends, which amounted to half of the net operating income. They no longer had an active part in farming. In the Vhembe District, two smallholder irrigation schemes were revitalised in this way, namely Makuleke and Block 1A of the Tshiombo scheme. The others remained unaffected. With reference to the use of micro-irrigation on smallholder irrigation schemes in South Africa, in 2011 there were 20 such schemes, 11 operational and nine non-operational. Combined they had a command area of 3 830 ha, 8.0% of the total.

3. Performance of smallholder irrigation schemes in South Africa

Globally, assessment of the performance of irrigated agriculture has received considerable attention, not in the least because of growing competition for water from other sectors (Faurès et al., 2007). Molden et al. (1998) developed a set of nine indicators to enable comparison of irrigation performance across irrigation systems. These covered irrigated agricultural output, water supply and financial returns. However, for smallholder irrigation schemes in South Africa the data required to calculate the nine indicators are rarely available. Most investigations into the performance of South African smallholder irrigation schemes used operational status, condition of the irrigation system, observations of cropping intensity and farm income in selected instances for assessment purposes. Generally, the conclusion of these studies has been that the contribution of smallholder irrigation schemes to social and economic development of irrigation communities has been far below expectations. (Bembridge & Sebotja, 1992; Bembridge, 1997; Bembridge, 2000; Machete et al., 2004; Tlou et al., 2006; Fanadzo et al., 2010). However, against a background of poor performance of smallholder irrigation schemes, few if any of the studies attempted to identify factors that appeared to contribute to differences in performance among these schemes. Such information could assist effective location and design of new schemes and also suggest priorities when planning the revitalisation of existing schemes.

4. Materials and methods

The Vhembe District is located in the Limpopo Province of South Africa (Fig.1), and is the most northern district of the Limpopo Province (Fig.2).

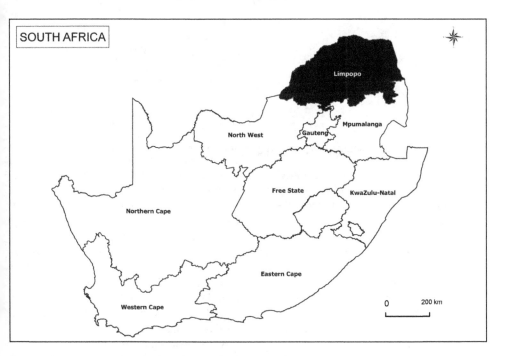

Fig. 1. The Limpopo Province in the north of South Africa

Vhembe borders Zimbabwe in the north and Mozambique in the east. It incorporates the territories of two former homelands, namely Venda and Gazankulu. The Venda homeland was created for the Venda-speaking people. Gazankulu was the territory allocated to the Tsonga-speaking people, also known as the Shangaan. Culturally, the BaVenda are closely associated with the Shona people of Zimbabwe, whilst the cultural roots of the Shangaan are in Mozambique.

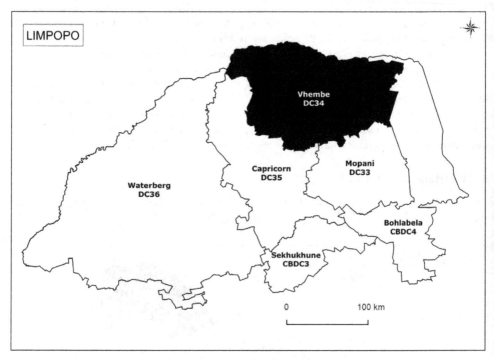

Fig. 2. Location of the Vhembe District in the Limpopo Province

Smallholder irrigation schemes in the Vhembe District were studied by means of a census. The census covered all smallholder irrigation schemes contained in the Vhembe register of the Limpopo Department of Agriculture, which was used as the sampling frame. A structured interview schedule was compiled for use as the survey instrument. The survey was conducted over a period of 10 months and involved four visits to the study area, each lasting between five and ten days. Work started in November 2008 and the last schemes on the list were visited during August 2009. Subsequently, the field data were scrutinised to identify data that were missing or needed verification. All the data queries that were identified were resolved during a follow-up visit to the study area in November 2009. Care was taken to achieve the greatest possible degree of reliability. Where possible, a small panel consisting of farmers, preferably members of the scheme management, and the extension officer were interviewed. At a few schemes only the extension officer or only farmers participated in the interview. Following the completion of the interview, a transect walk of the scheme was done and pictures were taken of selected features. A total of 42 schemes were identified but data collection at the Tshiombo Irrigation Scheme was done for each of the seven sub-units because of important differences amongst them. All other schemes were not subdivided, even when they consisted of multiple hydraulic units, referred to as irrigation blocks.

The first level of analysis was aimed at describing the population of smallholder irrigation schemes in the study area. For this analysis, the population was described one variable at a time, using descriptive statistics to generate summaries. The results provided a useful indication of the issues that affected smallholder irrigation in the study area and the diversity that surrounded these issues. The second level of analysis involved the testing of associations between four variables that were selected as performance indicators and a selection of independent variables that described the schemes.

The four performance indicators were operational status, number of years the scheme had been in operation, cropping intensity and degree of commercialisation. Operational status was selected because it is the primary indicator of performance. Once a scheme has stopped to operate, land use reverts to dryland agriculture. The number of years a scheme had been in operation was selected as an indicator of the durability of the system, which, in turn affects the rate of return on investment. Cropping intensity is a widely used indicator of the intensity with which water and land is being used in irrigated agriculture (Molden et al. 2007). Degree of commercialisation was selected because commercialisation has been shown to increase production and accelerate linkages in smallholder agriculture (Makhura et al., 1998).

The scheme characteristics that were considered in the analysis are shown in Table 1.

Scheme characteristic	Ranking criteria	Performance indicator			
		Operational status (n=48)	No of years in operation (n=48)	Cropping intensity (n=35)	Degree of commercialisation (n=35)
Hydraulic head	1 = gravity; 2 = pumped	Yes	Yes	Yes	Yes
Irrigation method	1 = surface; 2 = overhead; 3 = micro	Yes	Yes	Yes	Yes
Scheme area	Command area (ha)	Yes	Yes	No	No
No of plot holders	Population count	No	Yes	No	No
Plot size	Plot size (ha)	No	Yes	Yes	Yes
Organisation of production	1= individual; 2 = group	No	Yes	No	No
Water restrictions at scheme level	1 = no restrictions; 2 = seasonal restrictions; 3 = perpetual restrictions	No	No	Yes	No
Cash based land exchanges	0 = not practised; 1 = practised	No	No	Yes	Yes
Water theft	0 = not practised; 1 = practised	No	No	Yes	No
Effectiveness of scheme fence	1= effective; 2 = partially effective; 3 = not effective	No	No	Yes	No
Distance to urban centre	Distance by road (km)	No	No	Yes	Yes

Table 1. Selected characteristics of smallholder irrigation schemes in Vhembe, their ranking and the association of the performance indicators they were tested for

Inclusion of the scheme characteristics shown in Table 1 was justified as follows: hydraulic head for its direct effect on operational costs; irrigation method as an indicator of modernisation; scheme area and number of plot holders as indicators of management complexity; plot size for its association with degree of commercialisation identified in other studies (Van Averbeke et al., 1998; Bembridge, 2000; Machete et al., 2004); organisation of production, because group-based land reform projects have been shown to be prone to failure (Umhlaba, 2009); water restrictions because water is a production factor (Perry & Narayanamurthy, 1998); cash based land exchanges as an indicator of social and institutional responsiveness to demand for land (Shah et al., 2002); water theft as an indicator of social order (Letsoalo & van Averbeke, 2006); effectiveness of the scheme fence as a recurrent constraint in smallholder agriculture; and distance to urban centre as a measure of access to sizeable produce markets. Associations between scheme performance and scheme characteristics were assessed using Spearman's rank correlation. Scheme characteristics, their ranking and the association of the performance indicators they were tested for, are shown in Table 1. All 48 schemes were included in the analysis of operational status and number of years the scheme had been in operation. Schemes that were not operational (11), as well as two schemes that were no longer managed by plot holders following their revitalisation were excluded from the analysis of cropping intensity and degree of commercialisation.

5. Results

5.1 Summary description of smallholder irrigation schemes in Vhembe

Selected characteristics of the smallholder irrigation schemes in Vhembe District are presented in Table 2. Keeping in mind that the seven hydraulic units of Tshiombo were treated as separate schemes, 37 (77%) of the 48 smallholder schemes were operational and 11 were not. The smallest among the 48 schemes, Klein Tshipise, had a command area of only 8.5 ha, whilst the largest, Tshiombo, had a command area of 847 ha when its seven sub-units were combined. Together, the 48 schemes covered a total command area of 3760.1 ha, of which 3012.4 ha (80%) was located on schemes that were operational. The actual irrigated area on the schemes that were operational was 2693.1 ha. Two reasons were identified for the difference of 319.3 ha between command area and actual irrigated area on operational schemes. The first was infrastructural malfunctioning, which resulted in parts of the command area being withdrawn from irrigation. Schemes affected and areas involved were Khumbe (59 ha), Dopeni (17 ha) and Xigalo (30 ha). The second was that during revitalisation, parts of the command area were excluded, as in the case of Tshiombo Block 1A (8 ha) and Makuleke (204 ha).

At Tshiombo Block 1A (Fig.3), which was converted from canal to floppy irrigation, various small parts of the command area were not used because they did not fit the layout of the new irrigation system. At Makuleke, centre pivots limited use of the command area to selected parts of the scheme that were sufficiently large and homogeneous to accommodate a centre pivot. The food plot section of the scheme was never revitalised and remained non-operational at the time of the survey. Palmaryville lost 1.3 ha, when the demonstration plot was privatised.

5.2 Irrigation scheme development in Vhembe

The post-WWII period up to 1969 was very important for smallholder irrigation scheme development in Vhembe. Seven schemes with a total command area of 659.6 ha were

Scheme name	Operational	Number of years operational	Command area (ha)	Number of plot holders	Average plot size (ha)	Hydraulic head	Irrigation method
Nesengani	Yes	42	13.7	28	0.415	Pumped	Surface
Nesengani B1	No	17	20.6	116	0.178	Pumped	Overhead
Nesengani B2	No	17	40.9	116	0.352	Pumped	Overhead
Nesengani C	No	17	31.2	131	0.238	Pumped	Overhead
Dzindi	Yes	56	136.2	102	1.285	Gravity	Surface
Khumbe	Yes	56	145.0	138	0.623	Gravity	Surface
Dzwerani	No	20	124.0	248	0.500	Pumped	Overhead
Palmaryville	Yes	59	92.0	70	1.296	Gravity	Surface
Lwamondo	No	6	15.0	75	0.200	Pumped	Micro
Mauluma	Yes	45	38.0	30	1.267	Gravity	Surface
Mavhunga	Yes	45	47.5	32	1.532	Gravity	Surface
Raliphaswa	Yes	46	15.0	13	1.154	Gravity	Surface
Mandiwana	Yes	46	67.0	40	1.675	Gravity	Surface
Mamuhohi	Yes	46	77.0	61	1.262	Gravity	Surface
Mphaila	Yes	21	70.6	59	1.197	Pumped	Overhead
Luvhada	Yes	58	28.8	79	0.365	Gravity	Surface
Rabali	Yes	59	87.0	68	1.279	Gravity	Surface
Mphepu	Yes	49	132.8	133	0.998	Gravity	Surface
Tshiombo 1	Yes	48	60.5	47	1.287	Gravity	Surface
Tshiombo 1a	Yes	1	128.6	100	1.286	Pumped	Overhead
Tshiombo 1b	Yes	45	122.0	115	1.061	Gravity	Surface
Tshiombo 2	Yes	46	126.0	98	1.286	Gravity	Surface
Tshiombo 2a	Yes	48	173.5	114	1.522	Gravity	Surface
Tshiombo 3	Yes	45	128.4	100	1.286	Gravity	Surface
Tshiombo 4	Yes	46	56.0	112	0.500	Gravity	Surface
Lambani	No	4	260.0	16	16.250	Pumped	Surface
Phaswana	No	8	16.7	16	1.044	Pumped	Surface
Cordon A	Yes	45	43.7	38	1.150	Gravity	Surface
Cordon B	Yes	45	82.3	65	1.266	Gravity	Surface
Phadzima	Yes	45	102.3	103	0.993	Gravity	Surface
Makuleke	Yes	2	37.3	29	1.286	Pumped	Overhead
Rambuda	Yes	58	170.0	132	1.288	Gravity	Surface
Murara	Yes	42	70.0	7	10.000	Gravity	Surface
Dopeni	Yes	46	30.0	6	5.000	Gravity	Surface
Makhonde	No	10	83.0	58	1.431	Pumped	Micro
Sanari	No	17	17.0	11	1.870	Pumped	Micro
Tshikonelo	No	14	10.0	15	0.670	Pumped	Overhead
Chivirikani	Yes	28	68.3	112	0.609	Pumped	Overhead
Gonani	Yes	13	8.5	30	0.295	Pumped	Overhead
Folovhodwe	Yes	54	70.0	24	2.197	Gravity	Surface
Klein Tshipise	Yes	36	60.0	60	1.000	Gravity	Surface
Morgan	Yes	40	56.7	35	1.620	Gravity	Surface
Makumeke	Yes	5	17.0	63	0.269	Pumped	Micro
Dovheni	Yes	11	60.0	14	2.143	Pumped	Overhead
Mangondi	No	15	48.0	38	1.260	Pumped	Micro
Xigalo	Yes	5	22.0	24	1.080	Pumped	Micro
Garside	Yes	45	13.7	28	0.415	Gravity	Surface
Malavuwe	Yes	19	20.6	116	0.178	Pumped	Overhead

Table 2. Selected characteristics of the population of smallholder irrigation schemes in Vhembe District

Fig. 3. Revitalisation of Tshiombo Block 1A replaced the secondary canals and concrete furrows that conveyed water to the edge of individual farmers' fields with a centrally operated floppy sprinkler system that covers the entire hydraulic unit

established between 1951 and 1959. An additional 21 schemes with a total command area of 1978 ha were constructed during the decade that followed. This means that 2637.6 ha (70% of the existing smallholder irrigation scheme command area in Vhembe) were established between 1951 and 1969. All of the schemes that were constructed during this period were canal schemes. Nesengani, established in 1968, was the only canal scheme that made use of a pump to extract water to small concrete reservoirs from where it was gravitated to the plots. All other canal schemes extracted water by means of a weir or by means of spring diversion and relied entirely on gravity to convey water to the plots. All but one of the schemes that were constructed as canal schemes remained operational as canal scheme in 2009 but several had been fully or partially refurbished. The only exception was Block 1A of Tshiombo, which was recently (2008-09) transformed into a floppy irrigation scheme.

The period 1970 to 1979, which saw the construction of the last two canal schemes in Vhembe, namely Morgan and Klein Tshipise (Fig.4) in 1974, was a quiet period for smallholder irrigation development. Renewed activity occurred from 1980 onwards and was associated with the commencement of homeland self-government (Beinart 2001). All smallholder irrigation schemes that were established from 1980 onwards used pumps and pressurised irrigation systems. Dzwerani, established in 1980, was the first pressurised smallholder irrigation scheme in Vhembe. Non-operational at present, the 128 ha scheme at Dzwerani involved the pumping of water from the Dzondo River close to the confluence with the Dzindi River and the application of water by means of dragline sprinklers. Dzwerani was unique in that the 0.5 ha irrigation plots were also used for residential purposes. The idea for this development followed a visit by President Mphephu of the Venda homeland to Israel, where he observed similar arrangements. Dzwerani became a presidential pet project that received full financial support towards the cost of pumping and also towards other inputs, resulting in the development of a high degree of dependency on the state among the plot holder community. The project stopped operating when the pump was washed away during the 2000 flood. During the period 1980 to 1989, 10 of the existing schemes came into being with a combined command area of 495.8 ha. An additional 8 schemes were created between 1990 and 1999, with a combined command area of 506.7 ha. Most of these schemes were developed before 1998, when the homeland agricultural parastatals were still operational.

Fig. 4. The 8.5 ha Klein Tshipise scheme sourced its water from a spring and was the last canal scheme to be constructed in Vhembe, which occurred in 1974

During the first decade of the 21st century, the emphasis of state intervention was on revitalisation of existing schemes rather than on the creation of new schemes. Only two new schemes were established during this period, covering a modest command area of 120 ha. By the end of 2009, 10 of the 48 schemes covering a command area of 902.3 ha (24.0%) had been completely revitalised. Another 12 schemes covering a command area of 1083.3 ha (28.8%) had been partly revitalised and an additional two schemes with a combined command area of 61.5 ha (1.6%) were being revitalised. This brought the total number of schemes that had benefited from revitalisation, or would so soon, to 24, which was exactly half of the total. Combined the schemes that benefitted from revitalisation support covered 54.4% of the total smallholder scheme command area of 3760.1 ha, which left 1713 ha (45.6%) untouched.

5.3 Plot holder populations and plot sizes

Smallholder scheme land in Vhembe was held by a total of 3250 plot holders. Makhonde, with 7 plot holders was the scheme with the smallest plot holder population, whilst Tshiombo, when its seven sub-units were combined, had the largest with 660 plot holders. Dividing the total command area of the smallholder schemes by the total number of plot holders showed that on average, a Vhembe plot holder held 1.1570 ha of land, of which 0.8286 was operational irrigation land. However, plot size varied among the schemes. The smallest average plot size (0.178 ha) was found at Nesengani B1. Phaswana had the largest average plot size. The most common average plot size ranged between 1.01 ha and 1.5 ha and was found on 22 schemes. Plots in this size range were also dominant among the population of plot holders. A total of 1431 plot holders (44%) held plots that fitted in this size class.

5.4 Water sources, extraction and adequacy

Extraction of water directly from rivers using pumps or by means of weir diversion were the two most common ways in which smallholder schemes sourced their irrigation water. Spring water was used at two of the smaller schemes, namely Klein Tshipise (8.5 ha) and Luvhada (28.8 ha) and at Garside, spring water was used as a supplementary source. Makuleke was the only scheme that obtained its water from a dam. Surface irrigation, which invariably involved the use of short furrow irrigation (Fig.5), was dominant and occurred on 28 of the 48 schemes. All other methods of applying water were of secondary importance, perhaps with the exception of micro-irrigation (micro jet and drip), which was found on eight schemes but only two of these were operational in 2009.

Fig. 5. Short furrow irrigation at the Dzindi canal scheme

Generally, irrigation water availability was reasonably adequate, because only 5 of the 48 schemes reported year-round limitations, whilst on 21 schemes availability was said to be unlimited. Seasonal limitations in availability were mostly encountered on canal schemes. Four of the five schemes that reported availability to be always limited consisted of the last four irrigation blocks of Tshiombo, where lack of water was caused, at least in part, by the front-end blocks extracting more than their share, leaving too little for the tail-end blocks. Front-end tail-end differences in access to water among farmers were commonly reported on canal schemes. Mangondi, a drip irrigation scheme that was not operational in 2009, was the other scheme where water was always limited. Here the problem appeared not to be the source (Levhubu River) but rather the way the extraction system had been set up. Farmers used various ways of dealing with lack of irrigation water. In order of frequency of occurrence these included reducing the area planted to crops (53%), exchanging water among themselves (49%), stealing water from others (44%), reducing the frequency of irrigation (42%), irrigating at night (33%), planting crops that required less water (27%) and extracting water privately from the source using portable pumps (7%).

Only 27 schemes had a water license issued by the Depart of Water Affairs. Payment for water occurred at 17 schemes but water was paid for by the Limpopo Department of Agriculture, not the farmers. Water user associations had been established on 28 schemes, but with few exceptions these were not functional. Participation of scheme communities in catchment management activities was limited to a single case. On all but five schemes, management of water extraction and distribution was in the hands of an elected plot holder committee. At Tshiombo Block 1A and Makuleke, the commercial partner was in control and at Sanari (micro irrigation) and Dovheni (designed as a sprinkler line scheme), there was no management organisation and farmers were allowed to draw water whenever they wanted. At Phaswana, water management was the responsibility of the farmer cooperative, but the scheme was no longer operational.

Formal water management rules (captured in writing) had been drawn up at 37 schemes. At one scheme rules existed but had not been written up. The remaining schemes had no rule system in place to manage water. These included Lwamondo (collapsed micro irrigation scheme), Phaswana (non-operational micro irrigation scheme), Tshiombo Block 1A and Makuleke (revitalised schemes that had a commercial partner, who operated the scheme) and the pressurised schemes of Gonani and Dovheni, where water availability was said not to be limiting.

5.5 Land tenure and exchange

The Trust tenure system was by far the most prevalent tenure system on smallholder irrigation schemes in Vhembe. The implication was that land identified for the development of irrigation schemes had been detribalised and transferred to the state before the scheme was constructed. Trust tenure is regarded as the least secure of all systems that applied to African land holding and has been identified as a possible reason for the lack of land exchanges on smallholder irrigation schemes (Van Averbeke, 2008). Schemes with traditional tenure were usually established quite recently but there was one exception. Luvhada, a project developed in 1952 by the community of Mphaila without state assistance also had traditional tenure. Despite the prevailing Trust system of tenure, land exchanges occurred on 72% of the schemes, which was more common than expected. On schemes where land exchanges occurred the basis for the exchange in order of importance was cash (82%), free land preparation of own parcel (52%), a share of the crop (27%) and just as a favour (9%). The maximum duration of land exchange arrangements on schemes where such arrangements occurred was more than two years in 67% of the cases, up to two years in 12% of the cases and limited to a single season in 21% of the cases.

5.6 Farming systems, cropping intensity and degree of commercialisation

The most common farming system involved the production of grain (mostly maize) and vegetables. This farming system was found on 73% of the schemes. The crops that were incorporated in this farming system served both as food crops for own consumption and as crops than could be sold locally (Fig.6).

Fig. 6. The main farming system on smallholder irrigation schemes involved the production of maize and vegetables, both of which could be used for own consumption or sales

All other farming systems (primarily tropical fruit) were less important. In most cases they were established through the intervention of a homeland parastatal or through the implementation of the Joint Venture model. This model transferred control of the scheme to a commercial partner. In return for the release of their land for use by the Joint Venture, plot holders received dividends, which amounted to half of the net operating income. Cropping intensity varied considerably among the schemes. The majority of schemes had cropping intensities that ranged between 0.8 and 1.6. Schemes with cropping intensities higher than 1.2 were considered to be really active. The highest cropping intensity of 2.0 was found at Tshiombo Block 1A and Makuleke. Both were Joint Venture schemes where the commercial partner did all the farming. Across operational schemes, the proportion of produce that was sold was 50.6%, which was about 5% higher than the 45% recorded in the 1952 survey of smallholder irrigation schemes by the Tomlinson Commission (1955). The difference was partially due to the exceptionally high proportion of produce sold (99%) at the Makuleke and Tshiombo Block 1A Joint Venture schemes.

6. Performance assessment

The results of the statistical analysis of the association of the four performance indicators and selected characteristics of smallholder irrigation schemes in Vhembe are summarised in Table 3. All 48 schemes were included in the analysis of operational status and durability (number of years in operation). For the analysis of cropping intensity and degree of commercialisation, all non-operating schemes and the two schemes where a strategic partner was doing all the farming were excluded from the analysis.

Scheme characteristic	Performance indicator			
	Operational status (n=48)	No of years in operation (n=48)	Cropping intensity (n=35)	Degree of commercialisation (n=35)
Hydraulic head	-0.618 (0.000)	-0.848 (0.000)	0.057 (0.187)	0.270 (0.029)
Irrigation method	-0.707 (0.000)	-0.847 (0.000)	0.189 (0.071)	0.373 (0.007)
Scheme area	0.154 (0.074)	0.394 (0.002)	-	-
No of plot holders	-	0.348 (0.004)	-	-
Plot size	-	0.019 (0.225)	0.104 (0.104)	0.212 (0.055)
Organisation of production	-	-0.266 (0.018)	-	-
Water restrictions at scheme level	-	-	-0.438 (0.002)	-0.031 (0.215)
Cash based land exchanges	-	-	-0.014 (0.235)	-0.019 (0.229)
Water theft	-	-	-0.244 (0.041)	-
Effectiveness of scheme fence	-	-	-0.070 (0.174)	-
Distance to urban centre	-	-	0.067 (0.171)	-0.436 (0.002)

Table 3. Spearman's rank correlation coefficients and exact probabilities (bracketed) of the associations between four performance indicators and selected characteristics of smallholder irrigation schemes in Vhembe

6.1 Operational status

The correlation between the operational status of smallholder irrigation schemes in Vhembe and hydraulic head was fairly strong (-0.618) and statistically highly significant. The negative correlation coefficient indicated that gravity-fed schemes were more likely to be (and remain) operational than pumped schemes. The correlation between operational status and irrigation method was even stronger (-0.707). This suggested that schemes employing micro irrigation were less likely to be operational than schemes using overhead irrigation. Schemes using surface irrigation were most likely to be operational but this was to be expected because on all gravity-fed schemes plot holders made use of the short furrow method to apply irrigation water.

6.2 Durability

The number of years schemes had operated, or had been in operation, before they collapsed, was very strongly correlated with hydraulic head (-0.848) and irrigation method (0.847). The negative sign of both correlations indicated that canal schemes were considerably more durable than pumped schemes. The positive, statistically significant (p<0.01) correlation between scheme area and number of years in operation and between number of plot holders and number of years in operation were probably the result of the co-variation of these two factors with hydraulic head. On average, the plot holder population on gravity-fed schemes (71) was slightly larger than on pumped schemes (63) and the average scheme area of gravity-fed schemes (83.4 ha) was also larger than that of pumped schemes (55.7 ha). The statistically significant (p<0.05) negative correlation between organisation of production and number of years in operation indicated that group projects were less likely to last than projects where plot holders farmed individually. Plot size did not appear to affect durability of irrigation schemes.

6.3 Cropping intensity

Associations between cropping intensity and scheme characteristics were not very strong. Of all the scheme characteristics that were tested, cropping intensity was most strongly correlated with water restrictions at scheme level (r=-0.438). This negative correlation, which was statistically highly significant (p<0.01), indicated that water restrictions, mostly due to seasonal differences in the supply of the source, inhibited farmers from using their plots as intensively as possible. The weak but statistically significant (p<0.05) negative correlation between the occurrence of water theft and cropping intensity was probably the result of water restrictions causing water theft and not due to differences in the degree of social order at the schemes. Hydraulic head, effectiveness of the scheme fence and distance to the nearest urban centre appeared not to affect cropping intensity. Cropping intensity tended to be positively correlated with irrigation method (micro irrigation > overhead irrigation > surface irrigation) but the correlation was not statistically significant (p>0.05). Surprisingly, cropping intensity was not associated with the presence or absence of cash based land exchanges among plot holders.

6.4 Degree of commercialisation

Associations between degree of commercialisation and scheme characteristics were also not strong. Of all the scheme characteristics that were tested, degree of commercialisation was most strongly correlated with distance to the nearest urban centre (r=-0.436). The relative strength of this correlation indicated that access to local urban markets was a significant factor in determining the orientation of production of plot holders on smallholder irrigation schemes. Remoteness, which reduced access to markets, resulted in farmers focussing more on producing for own consumption. Hydraulic head (r=0.270) and method of irrigation (0.373) were positively correlated with degree of commercialisation. This indicated that plot holders on pumped schemes tended to orient their farming more towards markets than those on gravity-fed schemes and that commercialisation was stimulated by the use of overhead and micro irrigation. Degree of commercialisation tended to be correlated positively with plot size, but this correlation was not statistically significant (p>0.05). It needs pointing out that among the schemes that featured in the analysis of degree of commercialisation, the range in average plot size among schemes was limited, with the smallest plots being 0.295 ha and the largest 2.197 ha. Water restrictions and the prevalence of cash-based land exchanges did not appear to affect degree of commercialisation.

7. Discussion

In this study it was demonstrated that gravity-fed canal schemes, on which farmers practised short furrow irrigation, were more likely to be operational and to last longer than pumped schemes. This was in line with the observations of Crosby et al. (2000) and Shah et al. (2002) for South Africa at large. They pointed out that pumped schemes tended to offer better quality irrigation than gravity-fed schemes and that pumping costs helped to impose financial discipline. However, they also stated that pumped schemes were more vulnerable to breakdown and that the cost of pumping tended to squeeze the net operating income of farm enterprises. An analysis of smallholder irrigation projects in Kenya concluded that pumped schemes operated and maintained by groups of smallholders were not sustainable (Scheltema, 2002). All of these projects had collapsed even before it was time to replace the pump, because of their higher financial and organisation requirement relative to gravity-fed schemes. The current study also indicated that pumped schemes were more vulnerable to flood damage than gravity schemes, mainly because during heavy flooding, pumps were washed away.

Generally, the cropping intensity on the smallholder schemes in the study area was well below the optimum values of 1.5 to 2.5 suggested by Faurès et al. (2007) but higher than the cropping intensity of 0.45 recorded by Mnkeni et al. (2010) at the Zanyokwe smallholder irrigation scheme in the Eastern Cape. Under conditions of adequate water supply, the subtropical climate of Vhembe puts the achievement of cropping intensities of 2 and more within reach. The study showed that water restrictions were a significant factor in determining cropping intensity. The water restrictions encountered on schemes in Vhembe were mostly seasonal and were caused by fluctuations in supply at source, in line with the prevailing summer rainfall pattern. Reductions in cropping intensity in response to water restrictions were also observed by Perry & Narayanamurthy (1998) in Asia. The absence of any evidence of an association between cropping intensity and cash-based land exchanges (rentals) among plot holders contradicted the assertion of Tlou et al. (2006) that land tenure was the most important system on system factor in irrigated agriculture. Other researchers have also suggested that the development of land rentals would increase cropping intensity on smallholder irrigation schemes in South Africa (Shah et al., 2002; Van Averbeke, 2008) but the results obtained in this study did not support this anticipated effect.

Degree of commercialisation on smallholder irrigation schemes in Vhembe was found to be associated with the location of schemes in relation to local urban centres. As distance between scheme and urban centre increased, farmers were less likely to produce for marketing purposes. Van Averbeke (2008) reported that marketing of farmer's produce at the Dzindi Canal Scheme, which also formed part of the current study, was mainly in the hands of street traders. Street traders purchased fresh produce from farmers in small quantities on a daily basis and most of them (66 of 84) were sedentary traders, who retailed this produce to the public in areas characterised by heavy pedestrian flows, such as the main streets in towns and townships and at the entrance of hospitals. The other 18 street traders were mobile. They retailed produce in villages and townships that surrounded Dzindi, carrying a bag of produce on their heads as they moved from door to door. Bakkie traders, who purchased produce in larger quantities and transported this produce in their vans to the same type of trading places as those of street traders, also purchased produce at Dzindi, but relative to street traders, they were less important. Nearly all 66 sedentary traders who purchased fresh produce at Dzindi used combi-taxis to transport their produce to their retail places and 54 of the 66 used taxis to get to the urban centre of Thohoyandou where they sold the produce. Taxi fares between Dzindi and Thohoyandou were relatively cheap, because of the short distance. Therefore, the negative correlation between degree of commercialisation and distance between scheme and nearest urban centre suggests that the cost of taxi fares could well be the factor that determines whether or not it is financially viable for sedentary street traders to purchase from scheme farmers and travel to urban centres to sell. During the field work it was noted that when schemes were located far away from an urban centre, farmers mainly sold produce to residents around the scheme, to mobile street traders, who retailed door to door and to bakkie traders. The absence of sedentary street traders purchasing fresh produce in these remote schemes, contrary to Dzindi, is a plausible explanation for the negative correlation between degree of commercialisation and distance to urban centres, which was also reported by Magingxa et al. (2009) for a sample of smallholder schemes in various other parts of South Africa.

8. Conclusion

The study of factors affecting the performance of smallholder irrigation schemes in Vhembe District yielded several interesting results, which have implications for smallholder irrigation scheme policy. Smallholder canal schemes were more likely to be operational and to last longer than pumped schemes. This finding questions the desirability of converting canal schemes into pumped schemes, which has been the practice of the RESIS Recharge Programme of the Limpopo Department of Agriculture. The study results suggest that rehabilitating existing canal systems would most probably be more sustainable. The study also indicated that in the absence of external interventions, commercialisation among farmers on smallholder schemes was more likely to occur when schemes were located close to urban centres, because proximity made it financially viable for street traders to travel between scheme, as the place of purchase, and town, as the place of retail, using public transport. This is important when the development of new schemes is being considered. For remote schemes, external intervention aimed at supporting market access appeared to be necessary to enhance commercialisation. At this stage, few of the farmer-managed schemes received marketing support from external agencies. Efforts to that effect are recommended and should be facilitated by public extension services in collaboration with the private sector. Finally, the study results indicated that the two smallholder irrigation schemes that were consolidated and farmed as single entities by a strategic partner (commercial farmer) were characterised by high cropping intensities and high degrees of commercialisation. However, the sustainability of this revitalisation trajectory is highly questionable. The introduction of centre pivot or floppy systems largely prevent plot holders from repossessing their schemes and farms as individuals once the joint venture arrangement comes to an end. As was already pointed out by Crosby et al. (2000), 'the worst scenario (for smallholder irrigation scheme development) is where central management not only takes all decisions unilaterally on a top-down basis but also conducts all on-farm operations'.

9. Acknowledgement

Research towards this chapter was conducted as part of the solicited WRC Project K5/1464/4, entitled, "Best Management Practices for Small-Scale Subsistence Farming on Selected Schemes and Surrounding Areas through Participatory Adaptive Research", which was initiated, managed and funded by the Water Research Commission (WRC) and the non-solicited WRC Project K5/1804/4, entitled, "Improving Plot Holder Livelihood and Productivity on Smallholder Canal Irrigation Schemes in Limpopo Province", that was managed and funded by the WRC. The author wishes to thank Mr J Denison for his valuable contribution to the design of the survey instrument, Mr K Ralivhesa for his assistance during the fieldwork, Mrs L Morey of the Biometry Division of the Agricultural Research Council for statistical analysis of the data and my daughter Lerato Van Averbeke for useful suggestions on how to improve the readability of the text. The contributions to the study made by smallholders on irrigation schemes in Vhembe and by staff of the Department of Agriculture in the Vhembe District and the Thulamela, Makhado, Musina and Mutale Municipalities are also gratefully acknowledged.

10. References

Aliber, M., & Hart, T. (2009). Should Subsistence Agriculture Be Supported as a Strategy to Support Rural Food Insecurity? *Agrekon*, Vol.48, No.4, (December 2009), pp. 434-458, ISSN 0303-1853

Backeberg, G.R. (2006). Reform of User Charges, Market Pricing and Management of Water: Problem or Opportunity for Irrigated Agriculture. *Irrigation and Drainage*, Vol.55, pp.1-12, ISSN 1531-0361

Backeberg, G.R., & Groenewald, J.A. (1995). Lessons from the Economic History of Irrigation Development for Smallholder Settlement in South Africa. *Agrekon*, Vol.34, No.3, pp. 167-171, ISSN 0303-1853

Beinart, W. (2001*). Twentieth Century South Africa*. Oxford University Press, ISBN 0-19-289318-1, Oxford, UK

Beinart, W. (2003). *The Rise of Conservation in South Africa: Settlers, Livestock and the Environments 1770-1950*. Oxford University Press, ISBN 978-0-19-926151-2, Oxford, UK

Bembridge, T.J. (1997). Small-Scale Farmer Irrigation in South Africa: Implications for Extension. *South African Journal of Agricultural Extension*, Vol.26, pp. 71-81, ISSN 0301-603X

Bembridge, T.J. (2000). *Guidelines for Rehabilitation of Small-Scale Farmer Irrigation Schemes in South Africa*, WRC Report No 891/1/00, Water Research Commission, ISBN 1-86845-683-8, Gezina, South Africa

Bembridge, T.J., & Sebotja, I. (1992). A Comparative Evaluation of Aspects of the Human Impact of Three Irrigation Schemes in Lebowa. *South African Journal of Agricultural Extension*, Vol.21, pp.30-41, ISSN 0301-603X

Bundy, C. (1988). The Rise and Fall of the South African Peasantry, Second Edition, David Philip, Cape Town, South Africa

Christie, F., & Hanlon, J. (2001). *Mozambique & The Great Flood of 2000*, James Currey, ISBN 0-85255-857-0, Oxford, UK

Commission for the Socio-Economic Development of the Bantu Areas within the Union of South Africa (1955). *Summary of the Report*, The Government Printer, Pretoria, South Africa

Crosby, C.T., De Lange, M., Stimie, C.M., & Van der Stoep, I. (2000). *A Review of Planning and Design Procedures Applicable to Small-Scale Farmer Irrigation Projects*, WRC Report No 578/2/00, Water Research Commission, ISBN 1-86845-688-9, Gezina, South Africa

De Lange, M., Adendorff, J., & Crosby, C.T. (2000). *Developing Sustainable Small-Scale Farmer Irrigation in Poor Rural Communities: Guidelines and Check Lists for Trainers and Development Facilitators*, WRC Report No. 774/1/00. Water Research Commission, ISBN 1-86845-692-7, Gezina, South Africa

Denison, J., & Manona, S. (2007). *Principles, Approaches and Guidelines for the Participatory Revitalisation of Smallholder Irrigation Schemes: Volume 2 – Concepts and Cases*, WRC Report No TT 309/07, Water Research Commission, ISBN 978-1-77005-569-8, Gezina, South Africa

De Wet, C. (2011). Where Are They Now? Welfare, Development and Marginalization in a former Bantustan Settlement in the Eastern Cape, In: *Reforming Land and Resource Use in South Africa: Impact on Livelihoods*, P. Hebinck & C Shacleton, (Ed.), 294-314, Routledge, ISBN 978-0-415-58855-3, Milton Park, Abingdon, UK

Economic Development Department (23 November 2010). The New *Growth Path: The Framework*. Available from http://www.info.gov.za/view/DownloadFileAction?id=135748

Fanadzo, M., Chiduza, C., Mnkeni, P.N.S., Van der Stoep, I., & Stevens, J. (2009). Crop Production Management Practices as a Cause for Low Water Productivity at Zanyokwe Irrigation Scheme. *Water SA* Vol.36, No.1, (January 2010), pp. 27-36, ISSN 1816-7950 (On-line)

Faurès, J-M., Svendsen, M., & Turral, J. (2007). Reinventing irrigation, In: *Water for Food Water for Life: A Comprehensive Assessment of Water Management in Agriculture*, D. Molden (Ed.), 353-394, Earthscan, ISBN 978-1-84407-396-2, London, UK

Houghton, D.H. (1955). *Life in the Ciskei: A Summary of the Findings of the Keiskammahoek Rural Survey 1947-51*, SA Institute of Race Relations, Johannesburg, South Africa

Inocencio,A., Kikuchi, M., Tonosaki, M., Maruyama, A., Merrey, D., Sally, H., & De Jong, I. (2007). *Cost and Performance of Irrigation Projects: A Comparison of Sub-Sharan Africa and Other Developing Regions*. IWMI Research Report 109, International Water Management Institute, ISBN 978-92-9090-658-2, Colombo, Sri Lanka

Kamara, A., Van Koppen, B., & Magingxa, L. (2001). Economic Viability of Small-Scale Irrigation Systems in the Context of State Withdrawal: The Arabie Scheme in the Northern Province of South Africa, *Proceedings of the Second WARSFA/Waternet Symposium: Integrated Water Resource Management: Theory, Practice, Cases*, pp. 116-128, Cape Town, South Africa, October 30-31, 2001

Khandlhela, M., & May, J. (2006). Poverty, vulnerability and the impact of flooding in the Limpopo Province of South Africa. Natural Hazards Vol. 39, pp. 275-287, ISSN 1573-0840 (electronic)

Lahiff, E. (2000). *An Apartheid Oasis? Agriculture and Rural Livelihoods in Venda*, Frank Cass, ISBN 0-7146-5137-0, London, UK

Laker, M.C. (2004). *Development of a General Strategy for Optimizing the Efficient Use of Primary Water Resources for Effective Alleviation of Rural Poverty*, WRC Report No KV 149/04, Water Research Commission, ISBN 1-77005-208-9, Gezina, South Africa

Leibbrandt, M., & Sperber, F. (1997). Income and Economic Welfare, In: *From Reserve to Region - Apartheid and Social Change in the Keiskammahoek District of (former) Ciskei: 1950-1990*, C. De Wet & M. Whisson (Ed.), 111-152, Institute for Social and Economic Research, Rhodes University, ISBN 0-86810-331-4, Grahamstown, South Africa

Letsoalo, S.S., & Van Averbeke, W. (2006). Water Management on a Smallholder Canal Irrigation Scheme in South Africa. In: *Water governance for Sustainable Development: Approaches and Lessons from Developing and Transitional Countries*, S. Perret, S Farolfi, & R. Hassan (Ed.), 93-109, Earthscan, ISBN-13 978-1-84407-319-1, London, UK

Lewis, J. (1984) The Rise and Fall of the South African Peasantry: A Critique and Reassessment, *Journal of Southern African Studies*, Vol. 11, No. 1, (October 1984), pp. 1-24, ISSN 0305-7070

Lipton, M. (1996). Rural Reforms and Rural Livelihoods: The Context of International Experience, In: *Land, Labour and Livelihoods in Rural South Africa, Volume One: Western Cape*, M. Lipton, M. De Klerk & M. Lipton, (Ed.), 1-48, Indicator Press, ISBN 1-86840-234-7, Dalbridge, Durban, South Africa

Makhura, M.T., Goode, F.M., & Coetzee, G.K. (1998). A Cluster Analysis of Commercialisation of Farmers in Developing Rural Areas of South Africa. *Development Southern Africa*, Vol. 15, No. 3 (Spring 1998), pp. 429-445, ISSN 0376-835X

Machethe, C.L., Mollel, N.M., Ayisi, K., Mashatola, M.B., Anim, F.D.K., & Vanasche, F. 2004. *Smallholder Irrigation and Agricultural Development in the Olifants River Basin of Limpopo Province: Management Transfer, Productivity, Profitability and Food Security Issues*, WRC Report No: 1050/1/04, Water Research Commission, ISBN 1-77005-242-9, Gezina, South Africa

Magingxa, L.L., Alemu, Z.G., & Van Schalkwyk, H.D. (2009) Factors Influencing Access to Produce Markets for Smallholder Irrigators in South Africa. *Development Southern Africa*, Vol. 26, No.1 (March 2009), pp 47-58, ISSN 0376-835X

Mills, M.E.E., & Wilson, M. (1952). *Land Tenure*, Shuter and Shooter, Pietermaritzburg, South Africa

Mnkeni, P.N.S., Chiduza, C., Modi, A.T., Stevens, J.B., De Lange, M., Adendorff, J., & Dladla, R. (2010). *Best management Practices for Smallholder Farming on Two Irrigation Schemes in the Eastern Cape and KwaZulu-Natal Through Participatory Adaptive Research*. WRC Report No TT 478/10, Water Research Commission, ISBN 978-1-4312-0059-7, Gezina, South Africa

Molden, D., Sakthivadivel, R., Perry, C.J., De Fraiture, C., & Kloezen, W.H. (1998). *Indicators for Comparing Performance of Irrigated Agricultural Systems*, IWMI Research Report 20, International Water Management Institute, ISBN 92-9090356-2, Colombo, Sri Lanka

Scheltema, W. (2002). Smallholder Management of Irrigation in Kenya, In. *The Changing Face of Irrigation in Kenya: Opportunities for Anticipating Change in Eastern and Southern Africa*, H.G. Blank, C.M. Mutero & H. Murray-Rust, (Ed.), 171-189, International Water Management Institute, ISBN 92-9090-475-5, Colombo, Sri Lanka

Shah, T., Van Koppen, B., Merrey, D., De Lange, M., & Samad, M. (2002). *Institutional Alternatives in African Smallholder Irrigation: Lessons from International Experience with Irrigation Management Transfer Management Transfer*, IWMI Research Report 60, International Water Management Institute, ISBN 92-9090-481-X, Colombo, Sri Lanka

Stayt, H.A. (1968). *The BaVenda*, Second Impression, Frank Cass, London, UK

Tlou, T., Mosaka, D., Perret, S., Mullins, D., & Williams, C.J. (2006). Investigation of Different Farm Tenure Systems and Support Structure for Establishing Small-Scale Irrigation Farmers in Long Term Viable Conditions, WRC Report No 1353/1/06, Water Research Commission, ISBN 1-77005-475-8, Gezina, South Africa

Tomlison Commision (1955) see Commission for the Socio-Economic Development of the Bantu Areas within the Union of South Africa

Turral, H., Svendsen, M., & Faurès, J.M. (2010). Investing in Irrigation: Reviewing the Past and Looking to the Future. *Agricultural Water Management*, Vol. 97, pp. 551-560, ISSN 0378-3774

Umhlaba (2010). *A Review of Experiences of Establishing Emerging Farmers in South Africa: Case Lessons and Implications for Farmer Support within Land Reform Programmes*, Food and Agricultural Organization of the United Nations, ISBN 978-92-5- 106490-5, Rome, Italy

Van Averbeke, W. (2008). *Best Management Practices for Small-Scale Subsistence Farming on Selected Irrigation Schemes and Surrounding Areas through Participatory Adaptive Research in Limpopo Province*, WRC Report No TT 344/08, Water Research Commission, ISBN 978-1-77005-689-3, Gezina, South Africa

Van Averbeke, W., & Hebinck, P. (2007). Contemporary Livelihoods, In: *Livelihoods and Landscapes: The People of Guquka and Koloni and their Resources*, P. Hebinck & P.C. Lent, (Ed.), 285-306, Brill, ISBN 978-90-04-16169-6, Leiden, The Netherlands

Van Averbeke, W., M'Marete, C.K., Igodan, C.O., & Belete, A. (1998). *An Investigation into Food Plot Production at Irrigation Schemes in Central Eastern Cape*, WRC Report No 719/1/98, Water Research Commission, ISBN 1-86845-334-0, Gezina, South Africa

Van Rooyen, C.J., & Nene, S. (1996). What can we learn from Previous Small Farmer Development Strategies in South Africa? *Agrekon*, Vol. 35, No. 4 (December 1996), pp. 325-331, ISSN 0303-1853

Vermillion, D.L. (1997). *Impacts of Irrigation Management Transfer: A Review of the Evidence*, Research Report 11, International Irrigation Management Institute, ISBN 92-9090- 340-5, Colombo, Sri Lanka

Vink, N., & Kirsten, J. 2003. Agriculture in the National Economy, In: *The Challenge of Change: Agriculture, Land and the South African Economy*, L. Nieuwoudt & J Groenewald (Ed.), 3-19, University of Natal Press, ISBN 1-86914-032-X, Pietermaritzburg, South Africa

Vink, N., & Van Rooyen, J. (2009). *The Economic Performance of Agriculture in South Africa since 1994: Implications for Food Security*. Development Planning Division Working Paper No 17, Development Bank of Southern Africa, Halfway House, South Africa

9

Spatial Variability of Field Microtopography and Its Influence on Irrigation Performance

Meijian Bai, Di Xu, Yinong Li and Shaohui Zhang

National Center of Efficient Irrigation Engineering and Technology Research,
China Institute of Water Resources and Hydropower Research
China

1. Introduction

Surface irrigation is the main irrigation method,which is most widely used in the world.Surface irrigation performance is affected by field length,field width,field microtopography,inflow rate,soil infiltration,crops,and so on.Field microtopography is among the most important factors affecting the performance of basin irrigation system due to its influence on the advance and recession processes.It can direct the irrigation design and management to systematically analyze their effects on the basin irrigation performances by numerical simulation.

Field microtopography refers to the unevenness of a field surface,which may be characterized by a set of topographic data constituting the surface elevation differences (SED) between the actual and the target design elevations. The spatial variability of field microtopography includes a parameter characterizing the degree of unevenness and the spatial distribution of SED throughout the basin surface. The standard deviation (S_d) of SED is often used as an indicator of the degree of unevenness (Pereira and Trout 1999; Xu et al. 2002). However, for a same S_d the spatial distribution of SED may vary, which makes it difficult to describe the microtopography using a single indicator. Moreover, that single parameter does not allow to fully assessing the impacts of microtopography on the basin irrigation performances.

Because field-measured data are often limited and simulation modeling of surface irrigation is quite complex (*e.g.*, Walker and Skogerboe 1987; Strelkoff et al. 2000), studies on the influence of the microtopography on irrigation performances generally do not assume the spatial variability of SED, *e.g.*, Pereira et al. (2007). In fact, considering that spatial variability in modeling very much increases its complexity. However, not assuming the spatial variability of SED may lead to do not achieving an optimal solution for design and management of a basin irrigation system. Clemmens et al. (1999) and Li et al. (2001) generated basin surface elevations using a Monte-Carlo method basing upon the statistical characteristics of the surface elevation but they assumed that SED were random distributed inside a basin and did not consider its spatial variability.

Zapata and Playán (2000a) found that the spatial variability of surface elevations had much more influence on basin irrigation performance than the spatial variability of infiltration.

More attention should therefore be paid to the spatial variability of microtopography in basin irrigation design and evaluation to better support management and the decision making process relative to the target quality of precision leveling.

Considering the existing practical limitations in field microtopography characterization and in describing the impacts of the spatial variability of SED on basin irrigation performance, and aiming at supporting the improvement of basin irrigation systems in China, including the implementation of precision leveling, this study mainly comprises: (1) analyzing the semivariograms of SED for different basin types and the estimation of the semivariogram parameters from basin geometry parameters and the standard deviation of SED aiming at understanding the spatial dependence of surface elevations; (2) developing a stochastic model, adopting Monte-Carlo generation and kriging interpolation techniques, to generate SED data when knowing the respective standard deviation; (3) evaluating the influence of spatial variability of field microtopography on irrigation performance by numerical simulation.

2. Spatial variability of field microtopography

Based on the measured surface elevation data the spatial variability of field microtopography was analyzed using geostatistical technique.

2.1 Surface elevation difference (SED)

The surface elevation difference (SED) is defined as the difference between the observed and the target design elevations at each grid point i (z_i, cm), thus:

$$z_i = H_i - \overline{H_i} \tag{1}$$

where H_i is the observed elevation (cm) and $\overline{H_i}$ is the target design elevation(cm) at the same point i (i = 1, 2, ..., n). The degree of unevenness of SED is characterized by the standard deviation (S_d) of the z_i values:

$$S_d = \sqrt{\frac{\sum_{i=1}^{n}\left(z_i - \overline{z}\right)^2}{n-1}} \tag{2}$$

where $\overline{z}$ is the mean of SED (cm) observed in n grid points.

2.2 Geostatistics

The spatial variability of basin SED was analyzed using geostatistical techniques (Clark 1979). Experimental semivariograms $\gamma(h)$ were applied. These express the relation between the semivariance of the sample and the sampling distances:

$$\gamma(h) = \frac{1}{2N}\sum_{i=1}^{N}\left[z(x_i) - z(x_i + h)\right]^2 \tag{3}$$

where x_i is the coordinate of the observation point i; $z(x_i)$ is the respective SED value (m), h is the distance between pairs of observations (m), and N is the number of data pairs. The semivariograms models are defined with three parameters: the nugget (C_0), the sill (C_0+C), and the range (R). The nugget is the value of the semivariogram for a distance equal to zero. A non-null nugget may indicate either a systematic measurement error, or that a spatial variation occurs at a scale smaller than that used for measurements. The sill is the final stable value of the semivariogram. The range is the distance at which the semivariance reaches that stable value. As discussed by Barnes (1991), when the sample values are evenly distributed over an areal extent many times larger than the range of the variogram, then the sample variance is a reasonable first estimate for the variogram sill. When different conditions occur, the sample variance may, on the average, significantly underestimate the variogram sill. However, comparing the sample variance and the sill may be a good criterion for testing the validity of adopting a given experimental variogram model because if sill and variance differ greatly the experimental model is suspect (Barnes 1991).

The indicative goodness of fit (IGF) (Pannatier 1996) was adopted to quantify the fitting error when a theoretical semivariogram is adjusted to experimental data. The selected theoretical semivariogram is the one that produces minimal differences between observed and computed values. The IGF is given by

$$IGF = \sum_{i=0}^{n} \frac{P(i)}{\sum_{j=0}^{n} P(j)} \cdot \frac{D}{d'(i)} \cdot \left[\frac{\gamma(i) - \hat{\gamma}(i)}{\sigma^2} \right]^2 \tag{4}$$

where n is the number of lags, D is the maximum distance of lags, $P(i)$ is the number of pairs for lag i, $d'(i)$ is the distance for lag i; $\gamma(h)$ is the empirical semi-variogram for lag i; $\hat{\gamma}(i)$ is the theoretical semi-variogram for lag i; and σ is the standard deviation of analyzed data.

2.3 Basic data

Field-measured SED data were obtained through surveying of 116 basins located at Daxing and Changping in Beijing region, Xiongxian in Hebei Province, and Bojili in Shandong Province. Basin SED from Changping, Xiongxian and Bojili were observed using a topographic level with an accuracy of 1 mm at intervals of 5 to 10 m. The basin SED from Daxing was observed using both a topographic level and a GPS at intervals of 1.5 to 10 m; the accuracy of GPS was about 5 mm.

The observed basins were classified relative to their forms into three types depending upon the basin length (L) and width (W): strip basin, when the ratio $L/W > 3$ with $W \leq 10$ m; narrow basin when $L/W > 3$ with $W > 10$ m, and wide basin when $L/W < 3$. Table 1 summarizes related data on basins length, width, standard deviation of SED and average longitudinal slopes. It can be seen that the basins observed cover a large range of basin lengths, generally larger for the narrow basins. Basin widths also cover a large range; they are generally smaller in strip basins and larger in wide ones. S_d tends to be larger when the length is longer. The average longitudinal slope S_o is generally positive but small, not far from zero.

Basin parameters	Strip basins		Narrow basins		Wide basins	
	Range of observations	mean	Range of observations	mean	Range of observations	mean
Length (m)	30~278	84	50~300	158	20~200	93
Width (m)	1.9~10.0	4.9	10.0~35	19.0	10.0~80.0	51.0
S_d (cm)	0.80~4.50	1.93	1.20~5.30	3.11	1.50~4.00	2.53
Slope (‰)	0.1~4.3	1.0	0.0~3.6	0.9	0.0~3.3	1.1

Table 1. Main basin size and microtopographic parameters relative to the three basin types

2.4 Spatial structure of SED

The spatial structure of SED was analyzed using geostatistical techniques (see Section 2.2). Spherical semivariograms were fitted to the 116 observed basins. The descriptive statistics of the semivariogram parameters relative to the three basin types are presented in Table 2. Results show that the nugget is generally smaller for the strip basins and larger for the narrow ones. This may indicate that for narrow basins a spatial variation may occur at a scale smaller than that used for observations. The sill is also larger for the same basins. The range is not very different among the three types of basins and is larger when the basin length is longer. The ratio $C_0 / (C_0+C)$ averages 0.21, 0.34 and 0.32 respectively for strip, narrow and wide basins; these values indicate that a medium to strong spatial correlation exists for SED.

Three typical experimental semivariograms of SED having low, medium and high IGF are presented in Fig. 1. They refer to strip basins whose sizes are 30 × 6, 67× 2.5 and 82 × 7.5 m, respectively. They concern a spherical theoretical semivariogram, which is the one that best fitted the experimental data.

Basin type	Statistics	Semivariogram parameters				IGF
		Nugget (C_0) (cm²)	Sill (C_0+C) (cm²)	$C_0/(C_0+C)$	Range (R) (m)	
Strip basins	Maximum	2.20	22.00	0.67	60.00	0.097
	Minimum	0.00	0.80	0.00	5.00	0.02
	Mean	0.58	4.66	0.21	16.69	0.026
	Variance	0.52	0.47	0.27	0.47	0.60
Narrow basins	Maximum	8.0	29.00	0.67	58.00	0.071
	Minimum	0.00	1.45	0.00	6.00	0.003
	Mean	2.95	10.56	0.34	19.91	0.009
	Variance	0.63	0.62	0.64	0.54	0.49
Wide basins	Maximum	5.00	15.40	0.63	65.00	0.078
	Minimum	0.00	2.15	0.00	4.00	0.003
	Mean	1.92	6.89	0.32	25.83	0.012
	Variance	0.75	0.53	0.56	0.67	0.53

Table 2. Statistics of semivariogram parameters of SED for three basin types

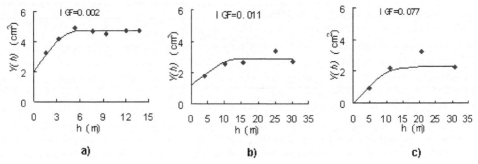

Fig. 1. The Experimental and Theoretical Semivariogram of SED for Different IGF

2.5 Calculation of the semivariogram parameters of SED

To search for the relationships among basin parameters and the parameters of SED semivariograms linear regressions were applied between every pair of parameters. Table 3 shows the correlation coefficients obtained and their significance level. Results show that the range mainly depends upon the basin length (L) and area (A), as well as on the observation distances (d). R also depends on the width (W) except for the strip basins which have a small W. The nugget is negatively correlated with the distance among observation points and shows relatively low correlation with the basin parameters; however, some significant relationship exists with the area and the length of the basins. The sill, as discussed before, is close to the variance of SED (S_d^2). The latter also relates to the range, mainly for the narrow basins. Based upon the relationships among basin parameters (L, A, S_d) and semivariogram parameters (C_0, C_0+C, R) empirical regression equations were established for the three types of basins (Table 4), which will be used for the developing of the stochastic modeling of field microtopography, for adjusting the generated SED in terms of spatial dependence of their values.

Basin type	Basin parameters	Nugget C_0	Sill C_0+C	Ratio $C_0/(C_0+C)$	Range R
Strip basins	L	-0.29	0.40**	-0.30	0.98**
	W	-0.28	0.21	-0.36*	0.05
	A	-0.34*	0.42**	-0.38*	0.90**
	S_d	-0.16	0.98**	-0.35*	0.39*
	d	-0.59*	0.31*	-0.39*	0.78**
Narrow basins	L	0.26	0.56**	-0.01	0.84**
	W	0.19	0.54**	-0.18	0.50**
	A	0.11	0.63**	-0.19	0.72**
	S_d	0.13	0.94**	-0.33	0.65**
	d	-0.69**	0.34*	-0.54**	0.67**
Wide basins	L	0.33*	0.21	0.05	0.89**
	W	0.25	0.22	0.01	0.91*
	A	0.24	0.16	0.01	0.93**
	S_d	0.17	0.93**	-0.43**	0.35*
	d	-0.70**	0.21	-0.31	0.87**

Note: * significance level 0.05; ** Significance level 0.01 L - length, W - width; A - area; S_d - standard deviation of SED; d - observation distances

Table 3. Coefficients of correlation relative to linear regressions between selected basin parameters and the parameters of the SED semivariograms for different basin types

Basin type	semivariogram parameters of SED		
	Nugget C_0 (cm²)	Sill (C_0+C) (cm²)	Range R (m)
Strip basin	$0.21S_d^2$	S_d^2	$0.18L+1.53$
Narrow basin	$0.34S_d^2$	S_d^2	$0.21L-4.11$
Wide basin	$0.32S_d^2$	S_d^2	$16.69A+5.26$

S_d^2 – variance of SED; L – basin length; A – basin area

Table 4. Empirical equations relating the parameters of the SED semivariograms with the basin characteristics for the three basin types

3. Stochastic modeling of field microtopography

3.1 Stochastic generation of SED

Considering both the randomness and the spatial dependence of basin SED values, the Monte-Carlo (M-C) method and kriging interpolation techniques were combined to develop a procedure for modeling microtopography. It consists of four steps:

1.*Stochastic generation of SED using the M-C method.* Based on the basins geometry (length L and width W), on the statistical characteristics of observed SED (mean $\bar{z}$ and standard deviation S_d), and on the observations grid spacings between rows (Δy) and columns (Δx), it is first determined the number n of elevation nodes to be randomly generated. Then n evenly distributed random numbers r_i [0, 1] are generated. The SED values z_i^0 corresponding to each r_i are computed through the following distribution:

$$F(z) = \int_{-\infty}^{z} \frac{1}{S_d \sqrt{2\pi}} \exp\left[-\frac{1}{2}\left(\frac{z-\bar{z}}{S_d} \right)^2 \right] d_z \tag{5}$$

where variables are those defined above. It results a set of generated SED values z_i^0 for all the grid nodes i.

2.*Adjusting the generated SED to the expected range of values.* In theory, the SED may assume any value but in practice their range is limited and depends upon the mean $\bar{z}$ and the standard deviation S_d that characterize each field. It was empirically assumed that the generated SED should fall within the interval [$\bar{z}$ -3S_d, $\bar{z}$ +3S_d]. Therefore, when any value z_i^0 is out of this range another value is generated using the M-C method.

3.*Establishing a spatial dependence for the generated SED values.* The generated SED values produce a spatial distribution different from the one of the actual microtopography that may be unrealistic because the proximity microtopographic relations between neighbor points are not considered. A kriging interpolation is then used to establish a spatial dependence of the generated SED values that considers the observed spatial structure of SED; New values for SED at each point i, z_i^1, are therefore estimated from the SED values at the neighbor points around i but assuming the above defined range of variation. Thus, the z_i^0 values are replaced by z_i^1 according to the relation

$$Z(z_i^1) = \sum_{j=1}^{M} \lambda_j (z_j^0) \tag{6}$$

where M is the number of points surrounding the point i, and λ_j are the weighing coefficients relative to the j neighbour points whose SED values are Z_j^0.

4.*Adjusting the generated SED for maintaining the original statistical characteristics.* The generation and adjustment procedures referred above cause that the statistical characteristics of SED are changed relative to the initial mean $\overline{z}$ and standard deviation S_d. Therefore, it is required to correct the generated SED aiming at assuring that their mean and standard deviation are conserved. First they are corrected for the mean and, afterwards, for the standard deviation, respectively:

$$z_i^2 = \frac{\overline{z}}{\overline{z}_1} z_i^1 \tag{7}$$

$$z_i^3 = (z_i^2 - \overline{z})\frac{S_d}{S_{d2}} + \overline{z} \tag{8}$$

where z_i^2 and z_i^3 are the values for SED after the respective corrections for the mean and the standard deviation, $\overline{z}_1$ is the mean of the z_i^1 values resulting from the kriging adjustment, and S_{d2} is the standard deviation of z_i^2 values.

3.2 Determining the number of SED generations

When SED are generated using the described stochastic modeling procedure, more than one set of SED can be generated for a given S_d. Different sets of SED generated with the same S_d will produce different values for the irrigation performance indicators when keeping constant all other factors that influence advance and recession. *i.e.*, the irrigation performance relative to a given SED set is unique. Thus, it is necessary to determine how many SED sets need to be generated for a given S_d to appropriately analyze the impacts of the spatial variability of the basin's microtopography on the irrigation performance.

3.2.1 Theoretical method

The number of SED generations can be determined by analyzing the trend of change of selected irrigation performance indicators resulting from the simulation of a given irrigation event through a number of SED sets. When N sets are generated for a given S_d then N sets of irrigation performance indicators are obtained by simulation of the same irrigation event. The number m ($m < N$) of SED generations required to characterize the population of basin SED may then be determined by analyzing the changes on irrigation performance with the number of SED generations.

Considering the population of an independent random variable X normally distributed with mean μ and variance σ^2, if $X_1, X_2, \ldots, X_m$ is a sample of size m from that population and whose mean is $\overline{X}$, then the probability for any value X_j ($j = 1, 2, \ldots, m$) to be included in the confidence interval of probability $1-\alpha$, is

$$P\left\{ \left|\frac{\overline{X}-\mu}{\sigma/\sqrt{m}}\right| < Z_{\alpha/2} \right\} = 1-\alpha \tag{9}$$

where $Z_{\alpha/2}$ is the value of the standard normal distribution corresponding to the probability $\alpha/2$. Therefore, the interval of estimation of the variables X_j (j = 1, 2, ..., m) relative to the same probability is $\left[\overline{X} - \dfrac{\sigma}{\sqrt{m}} Z_{\alpha/2}, \overline{X} + \dfrac{\sigma}{\sqrt{m}} Z_{\alpha/2}\right]$ (Mood et al. 1974; Deng 2002). Consequently, when aiming at an estimation precision l_0, the sample size required m shall satisfy the condition $\sigma \cdot Z_{\alpha/2} / \sqrt{m} \leq l_0$; thus, the number of SED generations, i.e., the sample size, should be at least

$$m = (\sigma \frac{Z_{\alpha/2}}{l_0})^2 \tag{10}$$

3.2.2 numerical experiment

A numerical experiment was developed to assess the number of sets of generated SED values for each basin type and S_d aiming at representing the possible land surface conditions to be analyzed through simulation for assessing the impacts of spatial variability of microtopography on basin irrigation performance.

Basin size and S_d were considered in numerical experiments to decide the number of SED generations. Data in Table 1 led to adopt as typical basin sizes 100 × 5 m, 150 × 20 m and 100 × 50 m respectively for the strip, narrow and wide basin types. For these typified basins, six degrees of surface unevenness are considered with S_d of 1, 2, 3, 4, 5, and 6 cm. Therefore, eighteen basin conditions resulted from combining those 3 basin sizes and the 6 S_d values. For each basin condition, 200 sets of SED were generated, thus producing 200 sets of irrigation performance indicators (water application efficiency (E_a),distribution uniformity(DU_{lq}) and average irrigation depth corresponding to the water justly cover the entire basin surface (Z_{adv})). In these simulations with the irrigation model B2D , the same soil infiltration parameters, Manning's hydraulics roughness n_r, soil water content when the irrigation starts and inflow rate conditions were adopted. The values for the Kostiakov-Lewis infiltration parameters and the Manning's roughness coefficient n_r were those obtained from a field experiment in a loamy soil in North China (k = 0.0045 m.min^{-a}, a = 0.46, f_0 = 0.0003 m.min^{-1}, n_r = 0.1 s.m$^{-1/3}$). The unit width inflow rate adopted was q = 4 l.s^{-1}.m^{-1}. The water cut-off time adopted was that required for the advance to be completed as practiced in North China, thus assuring that infiltration $Z >$ 0 in any point of the basin. The computational grid adopted was 1 × 1 m, 2 × 2 m and 5 × 5 m respectively for strip, narrow and wide basin types.

3.2.3 Setting the number of SED generations required for each basin type and S_d

Fig. 2 and Fig. 3, relative to a typical narrow basin, show results on the variation of the mean and the standard deviation of the performance indicators Z_{adv}, E_a and DU_{lq} with the number of SED generations (sample size). Results show that the mean and standard deviation of Z_{adv}, E_a and DU_{lq} do not change after a given threshold number of generations is reached, which depends upon S_d. Results for the wide and strip basins (not shown) are similar.

The mean values of the indicators Z_{adv}, E_a and DU_{lq} become nearly constant for a smaller number of generations than the respective standard deviation as shown in Figs. 2 and 3. Thus, the threshold number was computed from the latter, when the relative differences among six consecutive standard deviation values become smaller than 5%. The resulting values for the standard deviation of the referred indictors when they could be considered unchanged despite the number of generations increase are presented in Table 5 for the 3 basin types. Results show that those standard deviations are the smallest when $S_d = 1$ cm and increase with S_d. Greater changes occur for the strip basins because the length/width ratio is larger, which relate to its great influence on the advance process.

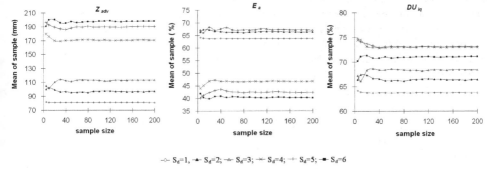

Fig. 2. Variation of the mean of the performance indicators Z_{adv}, E_a and DU_{lq} with the sample size (number of generated SED) for a typical narrow basin

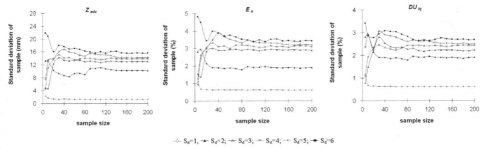

Fig. 3. Variation of the standard deviation of the performance indicators Z_{adv}, F_a and the DU_{lq} with the sample size (number of generated SED) for a typical narrow basin

Basin type	Performance indicator*	S_d=1 cm	S_d=2 cm	S_d=3 cm	S_d=4 cm	S_d=5 cm	S_d=6 cm
Strip basin	Z_{adv} (mm)	0.80	6.34	11.30	13.50	15.81	18.70
	E_a (%)	0.18	2.00	3.50	3.69	3.80	4.11
	DU_{lq} (%)	0.65	2.90	3.10	3.50	3.70	3.80
Narrow basin	Z_{adv} (mm)	1.26	10.21	13.01	13.80	14.23	15.59
	E_a (%)	0.62	1.89	2.90	3.10	3.20	3.42
	DU_{lq} (%)	0.62	1.89	2.21	2.40	2.50	2.69
Wide basin	Z_{adv}(mm)	1.05	8.44	10.25	12.37	15.37	17.02
	E_a (%)	0.66	2.56	2.81	3.00	3.42	3.56
	DU_{lq} (%)	0.65	2.35	2.49	2.79	2.87	3.06

* DU_{lq} - distribution uniformity, E_a – application efficiency, and Z_{adv} – infiltrated depth when the advance is completed

Table 5. Standard deviation of the irrigation performance indicators when their values become stable after simulating an irrigation event with a number of SED generations for various standard deviation (S_d) values .

The minimum number m of generations required for each basin type and various S_d values was computed with Equation 10 using variance data from simulations (Table 5). In this application the target precision l_0 are 3mm, 1% and 1% respectively for Z_{adv}, E_a and DU_{lq}, and the confidence level is associated with the probability $\alpha = 0.05$. Results for m are given in Table 6 showing that m increases with S_d, and are larger for Z_{adv} and smaller for DU_{lq}. Therefore, the number of generations adopted depends upon the indicator that is considered more important for the analysis. Because DU_{lq} is the best indicator of the system performance (Pereira et al. 2002), generally it is enough to consider the m values relative to this indicator. Otherwise, as for this study that pretends a wider analysis, the larger m value is selected, e.g., 33 SED generations would be required for the strip basin when $S_d = 2$ cm, and 55 when $S_d = 3$ cm.

| Basin type | Performance Indicator* | Precision l_0 | Number of SED generations | | | | | |
			$S_d = 1$ cm	$S_d = 2$ cm	$S_d = 3$ cm	$S_d = 4$ cm	$S_d = 5$ cm	$S_d = 6$ cm
Strip basin	Z_{adv}	3 mm	1	18	55	78	107	150
	E_a	1%	1	16	48	53	56	65
	DU_{lq}	1%	1	33	37	48	53	56
Narrow basin	Z_{adv}	3 mm	1	45	73	82	87	104
	E_a	1%	1	14	33	38	40	45
	DU_{lq}	1%	1	14	19	23	24	28
Wide basin	Z_{adv}	3 mm	1	31	45	66	101	124
	E_a	1%	1	26	31	35	45	49
	DU_{lq}	1%	1	22	24	31	32	37

* DU_{lq} - distribution uniformity, E_a – application efficiency, and Z_{adv} – infiltrated depth when the advance is completed

Table 6. Number of SED generations required for various standard deviation (S_d) values and basin types

3.3 Model validation

3.3.1 Field experiments for testing and validation of the stochastic model

Irrigation experiments were developed in a small level basin (30×15 m) located in the Experiment Station of the National Center of Efficient Irrigation Engineering and Technology Research, at Daxing, south of Beijing in the North China Plain. The soil was kept bare for easiness of observations. The soil texture is sandy loam and the average soil water content at field capacity and wilting point are respectively 0.26 and 0.10 m^3 m^{-3}. The basin was laser-controlled leveled. The observed standard deviation of SED is $S_d = 1.8$ cm.

The irrigation management followed the standard practice of winter wheat irrigation in this area. Different modes of water application into the basins were adopted: (1) fan inflow for the first irrigation, with the inflow point located by the middle of the upstream end of the basin; (2) corner inflow for the second irrigation, with the inflow concentrated at the upstream left corner of the basin; and (3) line inflow at the third irrigation, with water application at points distant 1 m along the upstream end of the basin. The water was conveyed to the field by a PVC pipe from the well pump where discharge was measured with a 1010WP-1/1010N supersonic flow meter. The average inflow rate was 12 1 .s⁻¹ for all irrigations. The water application was cut-off when the irrigation water covered the entire basin, i.e., the advance was completed.

A 1.5 ×1.5 m grid was used to perform all observations of soil surface elevation, and advance and recession times (Fig. 4). 12 measurement points were selected to obtain the cumulative infiltration curves before the first and second irrigation events (Fig. 4).

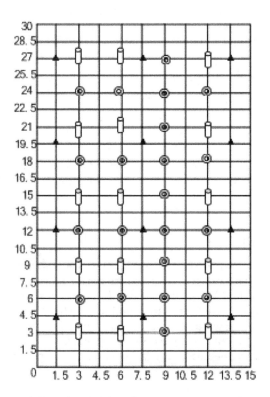

Fig. 4. Field measurements grid in the test basin: + advance and recession, surface elevation; ᕾ surface water depth; ᕾ and ◎ soil water content; ▲ soil infiltration.

The soil water content was observed one day before and after the irrigation to assess both the soil water deficit before irrigation and the infiltrated depth after it. Results were used to evaluate the irrigation performance. The soil water content was measured with a Time-Domain Reflectometry system type HH2 at 10, 20, 30, 40, 60 and 100 cm depth. Measurements were carried out at 36 grid points, i.e., adopting a 3 ×3 m grid (Fig. 4).

The surface water depth was measured using the water depth measuring device described by Li et al. (2006). This device is able to automatically measure and record the variation of water depth at given points during the whole duration of an irrigation event. Its testing results show that the adopted sensor is sensitive to the dynamic variations of the water depth with a precision of ±5 mm (Li et al. 2006). The water depth measuring devices were placed at every observation point before the irrigation starts and the recorded data was transferred to a computer after it ends. The measuring grid adopted is 6 × 6 or 6 × 3 m as described in Fig. 4.

3.3.2 Assessment indicators of model fitting

To assess the stochastic generation of SED, the irrigation model B2D was applied with both observed and generated SED data. Comparisons are made relative to the model computed irrigation performances. The average absolute error (AAE) and the average relative error (ARE) are used to assess the precision of simulations. These indicators are defined as:

$$AAE = \frac{1}{n}\sum_{i=1}^{n}|O_i - S_i| \tag{11}$$

$$ARE = \frac{100}{n}\sum_{i=1}^{n}\frac{|O_i - S_i|}{O_i} \tag{12}$$

where O_i and S_i are the values for the variables observed or simulated respectively, and n is the number of the observation points for the variables referred above. The subscripts OBS and GEN are used with these indicators when they result from simulations performed with measured and generated data, respectively.

3.3.3 results of model validation

The main observation data relative to the three irrigation events is summarized in Table 7. It can be observed that the first irrigation smoothed the basin surface, with S_d decreasing from 1.77 cm at the first event to 1.56 cm at the second one. Results show that adopting the traditional management, cutting-off the water application when the advance is completed, originates a non-uniform water application with very large differences between the maximum and minimum infiltration depths, 70 mm for the first event. Hence, the standard deviation of the infiltration depths (SD_Z) is large, 18 mm for that event. These non-uniformities produce low DU_{lq}.

Irrigation data (day/month)	SED Average (cm)	S_d (cm)	Inflow rate (l·s⁻¹·m⁻¹)	Irrigation time (min)	Infiltration depth Z(mm) Max	Min.	Average	Z_{lq}	SD_Z	Inflow type
26/11	5.09	1.77	0.8	41.3	103	33	66	46	18	Fan
15/4	4.76	1.56	0.8	38.1	84	36	61	44	13	Corner
20/5	4.93	1.57	0.8	40.0	92	28	64	37	15	Line

S_d – standard deviation of SED; SD_Z - standard deviation of infiltrated depths

Table 7. Selected results of irrigation experiments with different inflow types

Considering the basin size and the observed S_d, and taking into consideration the results in Table 6 when the analysis focus on all the indicators, 31 SED generations were performed for each observed S_d. The three irrigation events were then simulated with observed and generated SED data using the B2D model. The computational grid was 1.5 × 1.5 m; the infiltration data used were those observed in field experiments and the Manning's roughness coefficient was $n_r = 0.1$ s.m$^{-1/3}$ as indicated by Liu et al. (2003) for similar bare soil conditions. The resulting irrigation performance indicators (DU_{lq} and Z_{adv}) and advance time were used to compare the simulation results when observed or generated SED data (31 SED data sets) were input to the irrigation model.

Fig. 5 presents the advance curves observed and simulated with 5 minutes time steps referring to the 3 irrigation events, each one with a different inflow type (fan, corner and line). The simulated curves represented correspond to using as input the measured SED and a generated SED set which values are close to the average values.

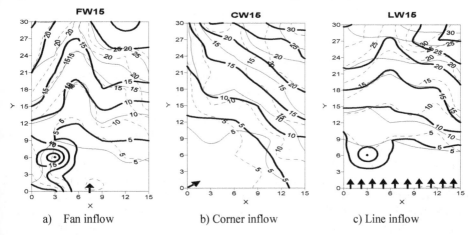

a)   Fan inflow   b) Corner inflow   c) Line inflow

Fig. 5. Advance curves observed (- - -) and simulated using measured SED (—) and generated SED data (—) with a time step of 5 minutes for a) fan inflow, b) corner inflow and c) line inflow.

The quality of these simulations is analyzed with the average absolute and relative errors (*AAE* and *ARE*) relative to all grid points for the case where observed SED are used, and the maximum, minimum and mean values of *AAE* and *ARE* relative to the 31 sets of generated SED (Table 8). The symbols OBS and GEN are used in this Table 8 to identify the simulations using observed and generated SED data.

Results show that differences between AAE_{OBS} and $AAE_{GEN,}$ as well as between ARE_{OBS} and $ARE_{GEN,}$ are small, *i.e.*, using generated SED data does not induce significant additional errors relative to using SED observed. However, the maximal errors are somewhat large but were infrequent. This means that using data generated with the same statistical characteristics as those observed in the field produce advance simulation results generally similar to those observed. Comparing the results relative to the three types of inflow into the basins (Table 8 and Fig. 5) it can be observed that line inflow is more accurately simulated by the B2D model than fan or corner inflow. Errors for the latter are the highest. This relates with the way how the water spreads from up- to downstream along the field and shows that when the inflow is concentrated the advance is more influenced by the microtopography of the basin surface. Results also show that the B2D model is an appropriate tool for 2-Dimension simulation of basins surface flow.

Inflow type	Irrigation event		AAE_{OBS} (min)	AAE_{GEN} (min)	ARE_{OBS} (%)	ARE_{GEN} (%)
Fan	1st	Mean	2.4	2.6	26.8	22.2
		Maximum		4.4		41.3
		Minimum		1.9		16.7
Corner	2nd	Mean	2.4	2.5	25.5	27.3
		Maximum		4.7		42.5
		Minimum		2.0		19.8
Line	3rd	Mean	1.9	1.7	18.7	13.8
		Maximum		3.8		35.1
		Minimum		1.4		12.6

Table 8. Average absolute and relative errors of estimation (AAE and ARE) of the advance time when simulations are performed with observed or generated SED

Table 9 presents the observed and simulated irrigation performances for fan, corner and line inflow. Because the irrigation cut-off was practiced when the advance is completed, for the three cases the distribution uniformity DU_{lq} is less good due to the small infiltrated depths downstream as shown in Table 7. Results are similar for the three inflow types. Table 10 presents the estimation errors for DU_{lq} and the infiltrated depth at time of cut-off, Z_{adv}. Data show that respective errors when using observed or generated SED data are similar and small, generally below 10%. Errors for the line inflow are smaller than for fan or corner inflow, which relates with results for advance referred before. Hence, it is possible to conclude that using various sets of generated SED data to analyze the impacts of basin microtopography on irrigation performances provides information similar to that derived from using observed SED values.

		Basin inflow type		
		Fan	Corner	Line
DU_{lq} (%)	From observations	70.1	72.6	57.9
	From measured SED data	64.1	65.3	60.1
	From generated SED data	62.9	65.3	62.1
Z_{adv} (mm)	From observations	66.0	61.0	64.0
	From measured SED data	69.2	63.5	62.8
	From generated SED data	63.0	57.4	60.7

* DU_{lq} - distribution uniformity, and Z_{adv} – infiltrated depth when the advance is completed

Table 9. Observed and simulated irrigation performance indicators using measured or generated SED data

Results above show that the stochastic modeling approach to generate the SED data allows a detailed study on impacts of microtopography on irrigation performance. Basin irrigation is applied in more than 95% of irrigated land in China, thus the improvement of these systems will have a great importance to overcome water scarcity and to provide for the sustainability of irrigated agriculture. As reported earlier, previous research has shown that land surface unevenness is a main factor contributing to low distribution uniformity and application efficiency in current basin irrigation systems. Various approaches are used for improving those systems including the use of modeling for design, such as the SIRMOD and SRFR models (Walker 1998; Strelkoff 1990) and the decision support system SADREG (Gonçalves and Pereira 2009). However, these models do not consider the effects of microtopography on the irrigation performance and it is advisable that their application follows a detailed study on such impacts that could provide for more realistic base assumptions for modeling. However, collecting field information on microtopography conditions is time and money consuming. Differently, adopting the approach developed in this study to generate a spatialized SED combined with the B2D model (Playán et al. 1994a, b) could be used to define the best improvement conditions for selected basin types predominant in various regions of China. Results for validation of the SED generation model shown above encourage its adoption in research practice oriented for surface irrigation improvement. This research is complemented with an evaluation of this modeling tool to assess the impacts of the spatial variability of mirotopography on the irrigation performance of various basins.

		Basin inflow type		
		Fan	Corner	Line
DU_{lq}	AAE_{OBS} (%)	6	7.3	2.2
	$Max\ AE_{GEN}$ (%)	10.6	10.6	6.1
	$Min\ AE_{GEN}$ (%)	3.8	4.6	2.9
	AAE_{GEN}(%)	7.2	7.3	4.2
	ARE_{OBS} (%)	8.5	10.1	3.8
	ARE_{GEN}(%)	9.5	9.4	5.3
Z_{adv}	AAE_{OBS} (mm)	3.2	2.5	1.2
	$Max\ AE_{GEN}$ (mm)	10	11	12
	$Min\ AE_{GEN}$ (mm)	0	0	0
	AAE_{GEN}(mm)	3	3.6	3.3
	ARE_{OBS} (%)	4.8	4.1	1.9
	ARE_{GEN}(%)	4.6	5.8	5.2

* DU_{lq} - distribution uniformity, and Z_{adv} – infiltrated depth when the advance is completed

Table 10. Absolute and relative errors of estimation (AAE and ARE) of the irrigation performance indicators when simulations are performed with observed or generated SED data

4. Influence of spatial variability of field microtopography on irrigation performances

4.1 Numerical experiments

Considering the statistical results relative to basin characteristics reported in Table 1 for 116 basins of North China, three representative basins were considered for the defined basin types strip, narrow and wide with sizes 100×5 m, 150×20 m and 100×50 m, respectively. For these basins, five degrees of surface unevenness were considered with S_d of 1, 2, 3, 4 and 5 cm. Two design slopes were adopted, $S_o = 0.1\%$ and zero leveled, as well as two inflow rates, $q = 2$ L.s^{-1}.m^{-1} and $q = 4$ L.s^{-1}.m^{-1}. For each basin type and S_d, the number m of SED generated with the generation model of spatial variability of microtopography (SVM model) is indicated in Table 11. The irrigation simulation model B2D was used to simulate the irrigation process relative to every SED data set. The basin irrigation performances referred above were computed for every simulation.

	$S_d = 1$ cm	$S_d = 2$ cm	$S_d = 3$ cm	$S_d = 4$ cm	$S_d = 5$ cm
Strip type	1	33	55	78	107
Narrow type	1	45	73	82	87
Wide type	1	31	45	66	101

Table 11. Number m of SED generations for each S_d and basin type

For the simulations with B2D, the same soil infiltration parameters, Manning's roughness n_r, initial soil water content and inflow conditions were adopted. Infiltration was characterized using the Kostiakov-Lewis equation. The infiltration parameters (k, a, f_0) and the Manning's roughness n_r were the same obtained in the field test in North China Plain used to validate the SVM model (See section 3.3.1) i.e., $k = 0.0045$ m.min^{-1}, $a = 0.46$, $f_0 = 0.0003$ m.min^{-1}, and $n_r = 0.1$ s.m$^{-1/3}$. This infiltration corresponds to a silty soil, whose layers are sandy loam or silt loam, and the average soil water content at field capacity and wilting point are respectively 0.26 and 0.10 m^3 .m^{-3}. Other simulation characteristics are the following: (a) the inflow time was the minimum irrigation time ensuring that advance could be completed, thus ensuring that the infiltration depth is $Z > 0$ everywhere in the basin; (b) the net target irrigation depth was set as $Z_{tg} = 80$ mm; (c) the inflow inlet was supposed to be located by the middle of the upstream end of the basin. According to the basin size and the simulation precision adopted, the calculation grids were 1×1 m, 2×2 m and 5×5 m, respectively for the strip, narrow and wide basins.

4.2 Irrigation performance indicators

The distribution uniformity of the low quarter, DU_{lq}, was selected as performance indicator in this study. It was defined (Merriam and Keller 1978) as:

$$DU_{lq} = 100 \frac{Z_{lq}}{Z_{avg}} \tag{13}$$

where Z_{lq} is the average low quarter infiltrated depth (mm) and Z_{avg} is the average depth of water applied to the field (mm).

In addition, the ratio

$$R_Z = Z_{adv} / Z_{tg} \tag{14}$$

between the average depth of water infiltrated following the complete advance criterion, Z_{adv}, (mm) and the net target irrigation depth, Z_{tg}, (mm) was used to assess the irrigation performance computed with the B2D model when simulating the irrigation events for the various SED generated sets. $R_Z > 1.0$ when overirrigation occurs, and $R_Z < 1.0$ when there is underirrigation. This indicator is used instead of the application efficiency because the latter is a management indicator that not only depends upon the variables characterizing the irrigation system but also upon the irrigator decisions, mainly referring to the timing of irrigation, that relates to the available soil water, and the depth applied, that determines the occurrence of deep percolation at a given irrigation event (Pereira 1999). Differently, R_Z indicates how the irrigation system is able to apply the target depth when influenced by land surface microtopography when the irrigation timing is appropriate. Z_{adv} was selected for the numerator of the ratio R_Z because Chinese farmers use to cut the inflow to the basins when the advance is to be completed, thus Z_{adv} indicates the expected infiltration when irrigation is managed that way.

4.3 Results and discussion

To characterize the influence of the spatial variability of microtopography on irrigation performance, simulations were performed for various S_d values (from 1 to 5 cm) and generating the number of variable SED referred in Table 11. Results in Fig. 6 show that infiltration at completion of advance, Z_{adv}, increases with S_d and, on the contrary, DU_{lq} decreases when S_d increases for zero-leveled basins ($S_o = 0$) but is insensitive to S_d for sloping basins ($S_o = 0.1\%$).

In sloping strip basins, when S_d increases from 1 to 5 cm, the average Z_{adv} increases from values close to the target $Z_{tg} = 80$ mm to values about 60% higher (Fig.6). If a zero leveled basin is considered, the average Z_{adv} becomes 80% higher, i.e. poorly leveled basins ($S_d \geq 4$ cm) .produce large overirrigation, mainly when no sloping. This reflects the role of the slope when the basin surface is uneven: advance is completed faster than for zero leveling. This also explains why farmers often adopted a mild slope and did not like to adopt zero leveling when improvements in surface irrigation were proposed (Cai et al. 1998). For these strip basins with slope, the average DU_{lq} shows little dependence on S_d but DU_{lq} increases when S_d decreases for zero leveled basins. The insensitiveness of sloping basins to S_d may be related to the fact that water keeps moving downwards after the advance is completed and is stored in the micro-depressions located downstream; therefore, infiltration is higher downstream, resulting that DU_{lq} in sloping strip basins cannot be high. In fact, it is limited to about 70%, thus indicating that an excellent performance is not achievable. Differently, for zero leveled basins a very high DU_{lq} is predicted when land leveling is excellent ($S_d = 1$ cm): 90% when the inflow rate is 4 L.s⁻¹.m⁻¹, and 86% when $q = 2$ L.s⁻¹.m⁻¹. These values progressively decrease when S_d increases; when $S_d = 5$ cm, DU_{lq} values are similar for zero leveled and sloping strip basins, near 70%.

For sloping narrow basins, the average Z_{adv} increases more than for strip basins when $S_d = 5$ cm and $q = 2$ L.s⁻¹.m⁻¹; for zero leveled basins and the same S_d and q, the average Z_{adv} increases to 186 mm, thus indicating an extremely high overirrigation. These differences in Z_{adv} relative to the strip basins are mainly due to the differences in length (100 vs. 150 m, respectively for the strip and narrow basins). Like for the strip basins, DU_{lq} is limited to about 74% and shows no sensitivity to changes in S_d, confirming that an excellent performance is not achievable with sloping basins. Differently, for zero leveled basins, very high average DU_{lq}, close to 90%, is predicted when leveling is excellent ($S_d = 1$ cm). For large S_d it results an average DU_{lq} close to that for sloping basins.

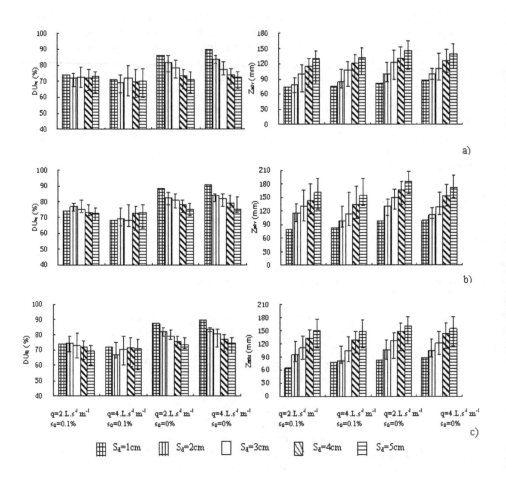

Fig. 6. Variability of the distribution uniformity DU_{lq}, and the infiltrated depth when the advance is completed Z_{adv}, as influenced by the microtopography (S_d varying from 1 to 5 cm) for basins with zero and 0.1% slope, and inflow discharges of 2 and 4 L s^{-1} .m^{-1}: a) strip, b) narrow, and c) wide basins (vertical bars indicate the range of variation for each case, the number of cases being that in Table 11)

For non-leveled (S_d = 5 cm) wide basins, the average Z_{adv} is 150 mm for q = 2 L.s^{-1}.m^{-1} and S_o = 0.1%, increasing to 162 mm when the slope is zero and adopting the same inflow discharge. For precision level basins and q = 2 L.s^{-1}.m^{-1}, the average Z_{adv} is much lower, 65 mm when S_o = 0.1% and 84 mm for zero leveling. These results are close to those for the strip basins; however Z_{adv} tends to be smaller for the latter. Differences to narrow basins relate to the large basin length of these ones, which produce a slower advance. Results for DU_{lq} are generally not far from those of strip basins. When S_o = 0.1% the average DU_{lq} is close to 72% and shows little dependence on S_d; for precision zero leveling DU_{lq} is close to 90% but decreasing to 74% when S_d = 5 cm.

Results above indicate that to achieve a high DU_{lq} zero leveling is required, preferably with a large inflow rate. When precise leveling is not applicable and water saving is intended, then a sloping surface is probably better for strip and narrow basins. If water saving is a priority, i.e., reducing Z_{adv}, it is required to adopt land leveling and a cutoff time smaller than the advance time. These results confirm those formerly obtained for strip basins in the North China Plain (Li and Calejo 1998) and long narrow basins in the lower reaches of the Yellow River (Fabião et al. 2003). Results for wide basins also identify the need for appropriate land leveling. Wide basins are adopted when paddy rice is in rotation with upland field crops. Since zero-level is the most adequate for paddy fields (Mao et al. 2004), it is interesting to have confirmed that zero-level is also the best option for higher performance of upland crops as defined in previous studies (Pereira et al. 2007).

An alternative way to appreciate the impacts of the spatial variability of microtopography on irrigation performance is to analyze the ratio R_Z (eq. 14) between Z_{adv} and Z_{tg} (Table 12). Results show that this ratio always increases when S_d increases, i.e., overirrigation increases with the basin surface unevenness. A value close to 1.0 indicates that a high DU_{lq} is achieved (cf. results in Fig. 6), also resulting in high potential application efficiency. If Z_{tg} would be larger, e.g. 100 mm, overirrigation would not occur for many cases with $S_d = 2$ cm and would be small with $S_d = 3$ cm. Ratios R_Z are smaller for the strip basins and larger for the narrow ones (Table 12). However, for the later the influence of the basin length is greater than that of the basin shape. Results for strip basins when $S_d \geq 3$ cm show that less overirrigation may be obtained for sloping fields and smaller inflow rates; for narrow and wide basins, better results for sloping basins refer to larger inflow rates. Differently, for precise leveled basins ($S_d \leq 2$ cm) the best results correspond to zero slope and large inflow discharges. These results justify the common option of farmers to apply large water depths (100 mm or more) and often adopting strip basins with lengths generally smaller than 100 m.

Basin type	Basin slope S_o (%)	Inflow rate (L s⁻¹ m⁻¹)	Z_{tg} = 80 mm S_d (cm)					Z_{tg} = 100 mm S_d (cm)				
			1	2	3	4	5	1	2	3	4	5
Strip basin	0.1	2	0.93	0.98	1.25	1.44	1.61	0.74	0.78	1.00	1.15	1.29
		4	0.93	1.04	1.34	1.53	1.64	0.75	0.83	1.07	1.22	1.31
	0	2	1.03	1.25	1.54	1.61	1.81	0.82	1.00	1.23	1.29	1.45
		4	1.10	1.24	1.39	1.58	1.74	0.88	0.99	1.11	1.26	1.39
Narrow basin	0.1	2	1.00	1.44	1.64	1.78	2.01	0.80	1.15	1.31	1.42	1.61
		4	1.04	1.21	1.42	1.69	1.93	0.83	0.97	1.13	1.35	1.55
	0	2	1.23	1.64	1.87	2.09	2.33	0.98	1.31	1.49	1.68	1.86
		4	1.25	1.40	1.62	1.92	2.15	1.00	1.12	1.29	1.53	1.72
Wide basin	0.1	2	0.81	1.18	1.38	1.65	1.88	0.65	0.94	1.11	1.32	1.50
		4	0.96	1.03	1.30	1.61	1.85	0.77	0.82	1.04	1.29	1.48
	0	2	1.04	1.33	1.58	1.85	2.03	0.84	1.06	1.26	1.48	1.62
		4	1.10	1.31	1.51	1.79	1.94	0.88	1.05	1.21	1.43	1.55

Table 12. Ratio between the infiltrated depth when the advance is completed, Z_{adv} (mm) and the target net depth (Z_{tg} = 80 and 100 mm) for various basin types, surface unevenness S_d (cm), basin slopes and inflow rates

5. Conclusion

Data on 116 basin irrigation fields, which cover a wide range of basin geometry and microtopography characteristics in various irrigation districts in North China were analyzed relative to the spatial variability of surface elevation differences (SED). The respective spatial structure is characterized with a spherical semivariogram model. Related data show that a medium or strong spatial dependence exist for the basins microtopography, and that a significant correlation exists between the semivariogram parameters and basin parameters (length, width, area, and the standard deviation S_d of SED).

Considering the characteristics of the spatial variability of SED, a procedure was developed for generating the spatial distribution of SED, and the number of SED generations required for each basin type and S_d was decided. field validation results showed that the stochastic tool developed for generating a spatial distribution of SED respecting a target mean and standard deviation is an useful research tool for a detailed analysis of to SED impacts on irrigation performance aimed at developing appropriate design criteria. These ones refer to land leveling, basin shape, basin lengths and inflow discharges.

Relative to leveling, if precision leveling is to be used, the standard deviation of surface elevation differences between the actual and the target design elevations should be $S_d < 2$ cm as already proposed in various studies. This threshold value should be used for both initial and maintenance land leveling. When this threshold is adopted both graded and zero leveled basins can be selected. Precise land leveling technology is available in China but due to the very intensive land use in North China, with wheat planted around five days after maize harvesting. and maize again following in the same land, very little time is left to perform precision leveling. Under these circumstances, it is acceptable to adopt graded basins as it is the general rule in China. When the slope is small such as analyzed herein (S_o = 0.1%) it results a distribution uniformity DU_{lq} smaller than for zero leveled basins but the excess infiltration relative to the target is smaller by about 20%. Thus, despite DU_{lq} is not maximized there are better chances for water saving. If uniformity is to be maximized, zero leveling is preferably adopted.

Following this study, it seems appropriate to adopt a decision support system and multicriteria analysis to better defining design options taking into consideration the costs and benefits associated with various possible alternatives, the expected impacts in water savings and the effects of uniformity of distribution on yields. These studies shall include different soils and infiltration characteristics as well as different basin sizes. A deeper understanding of economic, financial and environmental impacts is required to support developing appropriate design and issues for improving surface irrigation.

6. Acknowledgment

This research was supported by the National Natural Science Fund project No.50909100, the National High-Tech R&D Program Projects No. 2011AA100505. The collaborative Sino-Portuguese project on "Water Saving Irrigation: Technologies and Management" is also acknowledged. Thank Professor L.S. Pereira to give me good suggestions for this research work. Thanks are due to Zhang Shaohui and Li Fuxiang for contributing to the field experiments. The support by Dr. E. Playán on the use of the surface simulation model and further advising is sincerely acknowledged.

7. References

Barnes RJ (1991). The variogram sill and the sample variance. *Mathematical Geology* 23(4):673-678

Cai LG, Li YN, Liu Y, Qian YB (1998). Socio-economic aspects. Demonstration activities. In: Pereira, L.S. Liang, R.J., Musy, A. and Hann, M.(Eds.) *Water and Soil Management for Sustainable Agriculture in the North China Plain*. Dep. Engenharia Rural, Instituto Superior de Agronomia, Lisbon, pp. 382-405.

Clark I (1979) .Practical Geostatistics. Elsevier, London, 129 pp

Clemmens AJ, El-Haddad Z, Fangmeier DD, Osman HE-B (1999) .Statistical approach to incorporating the influence of land-grading precision on level-basin performance, *Trans. of ASAE*, 42(4): 1009-1017.

Clemmens AJ, Strelkoff TS, Playán E (2003). Field verification of two-dimensional surface irrigation model. *J Irrig Drain E-ASCE*, 129(6), 402-411

Deng HN (2002) .Method and Application of Probability Statistics. China Agriculture Press, Beijing (in Chinese).

Fabião MS, Gonçalves JM, Pereira LS, Campos AA, Liu Y, Li YN, Mao Z, Dong B (2003) .Water saving in the Yellow River Basin, China. 2. Assessing the potential for improving basin irrigation, Agricultural Engineering International Vol. V, LW 02 008. (http://www.cigrjournal.org/index.php/Ejounral/article/view/404).

Gonçalves JM, Pereira LS (2009). A decision support system for surface irrigation design. *J Irrig Drain E-ASCE*, 135(3): 343-356

Li YN, Calejo MJ (1998). Surface irrigation. In: Pereira LS, Liang RJ, Musy A, Hann M (eds) Water and Soil Management for Sustainable. *Agriculture in the North China Plain*. ISA, Lisbon, pp 236-303.

Li YN, Xu D, Li FX (2001). Modeling the influence of land leveling precision on basin irrigation performance. *Transactions of the CSAE*, 17(4):43-48 (in Chinese).

Li YN, Xu D, Li FX (2006). Development and performance measurement of water depth measuring device applied for surface irrigation. *Transaction of CSAE*, 22(1):32-36.

Liu Y, Cai JB, Li YN, Bai MJ (2003) .Assessment of crop irrigation requirements and improvements in surface irrigation in Bojili irrigation district. In: Pereira LS, Cai LG, Musy A, Minhas PS (eds), Water Savings in the Yellow River Basin. Issues and Decision Support Tools in Irrigation. China Agriculture Press, Beijing, pp 131-152.

Mao Z, Dong B, Pereira LS (2004) .Assessment and water saving issues for Ningxia paddies, upper Yellow River Basin. *Paddy and Water Environment* 2(2): 99-110.

Merriam JL, Keller J (1978) .Farm irrigation system evaluation: A guide for management. Agricultural and Irrigation Engineering Department, Utah State University, Logan, Utah, 271pp.

Mood AM, Graybill FA, Boes DC (1974) .Introduction to the theory of statistics. McGraw-Hill, New York, 564 pp.

Pannatier Y (1996) .VARIOWIN: Software for Spatial Data Analysis in 2D. Springer-Verlag, New York.

Pereira LS (1999). Higher performances through combined improvements in irrigation methods and scheduling: A discussion. *Agric. Water Manage*. 40 (2-3): 153-169.

Pereira LS, Gonçalves JM, Dong B, Mao Z, Fang SX (2007) .Assessing basin irrigation and scheduling strategies for saving irrigation water and controlling salinity in the Upper Yellow River Basin, China. *Agric. Water Manage*. 93(3): 109–122.

Pereira LS, Oweis T, Zairi A (2002). Irrigation management under water scarcity. *Agric. Water Manage.* 57: 175-206.

Playán E, Faci JM, Serreta A (1996a). Characterizing microtopographical effects on level-basin irrigation performance. *Agric. Water Manage.* 29:129-145.

Playán E, Faci JM, Serreta A (1996b). Modeling microtopography in basin irrigation. *J Irrig Drain E-ASCE*, 122(6): 339-347.

Playán E, Walker WR, Merkley GP (1994a). Two-dimensional simulation of basin irrigation: I: Theory. *J Irrig Drain Eng*, 120(5): 837-856.

Playán E, Walker WR, Merkley GP (1994b) .Two-dimensional simulation of basin irrigation: II: Applications. *J Irrig Drain Eng*, 120(5): 857-869.

Strelkoff TS (1990) .SRFR: A computer program for simulating flow in surface irrigation furrows-basins-borders. WCL Rep. 17, U.S. Water Conservation Laboratory, USDA/ARS, Phoenix.

Strelkoff TS, Clemmens AJ, Schmidt BV (2000) .ARS software for simulation and design of surface irrigation. In: Evans RG, Benham BL, Trooien TP (eds) Proceedings of the 4th Decennial National Irrigation Symposium, ASAE, St. Joseph, MI, pp. 290-297.

Walker WR (1998). SIRMOD – Surface Irrigation Modeling Software. Utah State University, Logan.

Walker WR, Skogerboe GV (1987). Surface Irrigation. Theory and Practice. Prentice-Hall, Englewood Cliffs, NJ.

Xu D, Li YN, Chen XJ, Xie CB, Liu QC, Huang B (2002). Study and Application of the Water Saving Irrigation Technology at Farm. China Agriculture Press, Beijing (in Chinese)

Xu D, Li YN, Li FX, Bai MJ (2005) .Analysis of feasible grid spacing in agricultural land levelling survey. *Transactions of the CSAE*, 21(2):51-55. (in Chinese)

Zapata N, Playán E (2000a) .Simulating elevation and infiltration in level-basin irrigation. J Irrig Drain Eng, 126(2): 78-84.

Zapata N, Playán E (2000b) .Elevation and infiltration in a level basin. I. Characterizing variability. *Irrigation Sci.*, 19: 155-164.

10

Decision Strategies for Soil Water Estimations in Soybean Crops Subjected to No-Tillage and Conventional Systems, in Brazil

Lucieta G. Martorano[1], Homero Bergamaschi[2],
Rogério T. de Faria[3] and Genei A. Dalmago[4]
[1]Embrapa Amazônia Oriental
Travessa Eneas Pinheiro s/n, Belém, PA
[2]UFRGS and CNPq, P. Alegre, RS
[3]FCAV-Unesp, Jaboticabal, SP
[4]Embrapa Trigo
Passo Fundo, RS
Brazil

1. Introduction

Conservationist practices have been increasingly adopted in Brazil during the last thirty years, especially with the change from the conventional cropping system to the no-tillage system. The latter has been widely spread in several Brazilian regions where the soybean crop takes part in annual crop rotation (Denardin et al., 2005). In the Center-Southern region it reached nearly 80% of grain producers. Among environmental and economic gains are: increasing crop yield, soil water and carbon stocks increments, reductions of production costs, control of soil erosion, mitigation of carbon emissions and water crop deficit. According to Buarque (2006) sustainable agriculture involves several structural changes and faces social and political resistance. In order to mitigate environmental impacts on food production and ecosystem services, policies should aim to develop more resilient cropping systems and provide sustainable management of natural resources.

Soybean is a commodity of great interest in national and international markets of which Brazil is the second largest producer in the World. Climate variability can severely affect crop yield and reduce total food production. Soybean production has proved to be highly dependent on climate to achieve genetic potential of the cultivars used by farmers. Among climate variables, studies in Rio Grande do Sul State, in Southern Brazil, showed that soybean production was mostly correlated with rainfall. Interannual variability in rainfall has been considered the main cause of fluctuations in grain yields (Bergamaschi et al., 2004) for non-irrigated crops.

Periods of water deficits from January to March are frequent and usually coincide with the summer crop critical period (flowering and grain filling), limiting the yield of soybean in the state (Matzenauer et al., 2003).

Decision support systems, such as DSSAT (Decision Support System for Agrotechnology Transfer) can be very useful for cropping system planning. The models included in that software were calibrated for different climate and soil conditions and crop management system and also applied by researchers worldwide (Jones et al., 2003). Different physical and physiological processes are simulated by the DSSAT models, such as photosynthesis, respiration, biomass accumulation and partitioning, phenology, growth of leaves, stems and roots (Hoogenboom, 2000, Hoogenboom et al., 2003), and soil water extraction (Faria & Bowen, 2003). After calibration at a site in the state of Rio Grande do Sul, Brazil, the model CROPGRO-Soybean showed a high performance to simulate grain yield and crop growth and variables of development under the no-tillage cropping system and the conventional system (Martorano, 2007; Martorano et al., 2007; Martorano et al., 2008a; Martorano et al., 2008b; Martorano et al., 2009). DSSAT is frequently updated (v4.5 is on http://www.icasa.net), and currently there is a group of researchers working on the calibration of DSSAT/ CENTURY in order to assist in decision strategies.

This chapter presents experimental results used in the assessment of the performance of the DSSAT models, carried out in a site in Rio Grande do Sul, Brazil. This study aimed at the establishment of scenarios for sustainable agriculture, based on principles of data interoperability and DSSAT users' network tools.

2. Problems and strategies for soybean crops

Several factors should be considered when evaluating low carbon agriculture, for example, the land use and agricultural management, the correct crop conduction, the evaluation of edapho-climatic characteristics, ethnic and cultural respect, and the aggregation of goods and services to the society. Some studies on sustainability indicators have pointed out that cropping systems, such as the no-tillage system, reduce environmental impacts, improve productivity and have lower production costs. Global climate projections show that temperature increases in some areas with high temperatures could worsen food production problems. Variation in productivity due to water availability is already a common problem for rain fed crops, which are the majority of crop areas in Brazil.

The Brazilian economy is closely linked to the supply of natural resources, especially water use, in agriculture, hydropower generation, industrial sectors, and other human demands. However, the lack of water management may cause several impacts and serious threats to the human population. It is known that water has plenty sources in many regions of Brazil, but this resource may become scarce if there is no concern regarding its maintenance.

Awareness of appropriate use in different segments of the productive sector is indispensable to enable quantifying the "water footprint" (Hoekstra & Chapagain, 2007), in these sectors. For agriculture, the concerns have turned toward waste of water, not only in arid regions of the world, but also the waste related to improper use and decisions on when, how and what is the most efficient technology to be applied in irrigated crops, for instance.

Decision support tools, such as DSSAT (Decision Support System for Agrotechnology Transfer), show high potential for decision makers to improve management of soybean crops in Brazil, after calibration of the models. Martorano (2007) and Martorano et al (2008a) showed that the CROPGRO-Soybean model had high performance simulating phenological phases, growth and yield under irrigated condition, both in conventional tillage and no tillage. DSSAT models can be a suitable tool to assess the effects of tillage on soil carbon in order to mitigate carbon emissions to the atmosphere (Martorano et al, 2008b). The model can simulate realistic scenarios for decision-makers (farmers, managers, agricultural technicians and government), as well as to identify crops problem for scientists defining research priorities.

3. Material and methodology

An example of research on soil water status is presented based on a field experiment. To evaluate soil-plant-atmosphere processes associated with soil management on the express condition of water, an experiment was conducted in the cropping season of 2003/04 with soybeans (cv. Fepagro RS10, long cycle), at the Experimental Station of the Federal University of Rio Grande do Sul State (EEA/UFRGS), in Eldorado do Sul, Brazil (30º 05'27"S; 51º 40'18" W, altitude 46m). The experiment was sown on 2003, Nov 20 in a typical dystrophic red clay soil, with plots conducted under no-tillage (NT) and conventional tillage (CT), irrigated (I) and not irrigated (NI).

Crop was sown 0.40 m between rows, with an average population of 300,000 plants ha⁻¹. An automatic meteorological station recorded weather variables and tensiometers (mercury-Hg column) measured daily soil water matric potential. Weekly assessment on plant growth and development were taken (Fig. 1). Leaf area index (LAI) and dry biomass (leaves, stems, pods and seeds) were determined weekly. Model input included minimum and maximum air temperature (°C), precipitation (mm) and solar radiation (MJ m⁻²). Soil inputs included soil classes, soil physical-hydraulic and chemical properties, in addition to crop management information (weed control, variety, planting date and irrigation).

Irrigation was applied when soil matric potential reached -60 kPa, as measured by tensiometers installed in irrigated no-tillage plots (NIT). The crop water use was monitored by a weighing lysimeter cultivated with soybean under conventional tillage. Each management system contained two batteries with tensiometers placed at depths (m) of 0.075, 0.15, 0.30, 0.45, 0.60, 0.75, 0.90 and 1.05 m and one with the same depths in addition to a tensiometer at 1.20m. Readings were made every day, around 9 p.m. (local time).

Fig. 1. Meteorological station (A), experimental design (B), soybean in no-tillage system (C), lysimeter (D), batteries with tensiometers (E) in experimental area at the EEA / UFRGS, 2003/04, in Rio Grande do Sul State, Brazil, and layout of software DSSAT initial page (F).

With matric potential values, the corresponding soil moisture was calculated using the soil retention curves obtained experimentally by Dalmago (2004) for no-tillage and conventional tillage plots. The program of Dourado Neto et al. (2005) was used to calculate the volumetric water content (Soil Water Retention Curve-SWRC, v.1.0), using Equation 1, Van Genuchten (1980):

$$\theta v = \theta r + \frac{(\theta s - \theta r)}{\left[(1 + \alpha \Psi m)^n\right]^m} \tag{1}$$

where the volumetric water content (cm^3 cm^{-3}) is represented by the θv and humidity and residual saturation θr and θs, respectively. The matric potential of soil water (kPa) is represented by Ψm and coefficients (dimensionless) by the letters α, n and m.

In Decision Support System for Agrotechnology Transfer was considered an experimental data of soil, climate and specifications of soil management. The model selected for the legume was CROPGRO Soybean. In this chapter, attention turned to the evaluation of model performance to simulate soil water content. CROPGRO-Soybean was calibrated, using genetic coefficients for the cultivar Fepagro RS-10, as described by Martorano (2007) and Martorano et al (2008a). The methodology of Willmott et al. (1985) recommended the use of RMSE (root mean square error) and D-index (index of agreement), but suggested that RMSE is the "best" measure as it summarizes the mean difference in the units of observed and predicted values (Martin et al., 2007). The RMSE indicates the bias produced by the model, i.e., deviation of the actual slope from the 1:1 line while it also may be seen as a precision measure, as it compares the bias of model predicted values with the random variation that may occur. The D-index is a descriptive (both relative and bounded) measure, and can be applied to make cross-comparisons between observed and simulated data (Loague & Green, 1991) by equation 2 and 3.

$$D = 1 - \left[\frac{\sum_{i-1}^{N} (Pi - Oi)^2}{\sum_{i-1}^{N} (|P'i| + |O'i|)^2} \right] \tag{2}$$

where N is the total of observations, Pi and Oi are respectively predicted and observed values , P'i refers to the absolute difference between Pi and the average of predicted variable P, and O'i is the difference between the observed value Oi and the average of observed variable O. The closer to unity is the D-index (Willmott et al., 1985), the higher the index of agreement between observed data and simulated value by the model is. Also, the observed data were compared with those simulated with the mean square error (RMSE), using equation 3.

$$RMSE = \sqrt{\frac{\sum_{i-1}^{N} (Pi - Oi)^2}{N}} X \frac{100}{M} \tag{3}$$

where Pi and Oi are the values of the variables simulated by the model and observed in the field, corresponding to the evaluation interval. The RSME expresses the relative difference

(%) between observed and simulated by CROPGRO-Soybean. It is considered highly accurate when the RMSE is less than 10%, good precision between 10% and 20%, and between 20% and 30%, which depending on the boundary condition may be acceptable. The model showed low accuracy when error is above 30%.

In this chapter, attention is given to evaluating the model performance for simulating the soil water content. The mean square error (RMSE) was used to expresses the relative difference (%) between the observed data and estimations by CROPGRO-Soybean, as described by Martorano (2007) and Martorano et al (2008a).

According to data from soil water content and grain yield was applied analysis of variance and means were compared using the Tukey test at 5% significance level.

4. Results and discussion

Evaluating the meteorological information during the soybean cycle (cv. Fepagro RS-10) showed that solar radiation ranged between 8.8 and 27.9 MJ m^{-2}day^{-1} (Fig. 2), and the average of 20.7 MJ m^{-2}day^{-1}, corroborate with the climate average for the period in the region which is about 20 MJ m^{-2}day^{-1} (Bergamaschi et al., 2003). There were two solar radiation peaks; one in January, the first ten-day period of (J$_1$), the order of 27 MJ m^{-2} day^{-1}, and another in March, the second ten-day period (M$_2$) close to 25 MJ m^{-2} days^{-1}. These values corroborate with those presented by Cargnelutti Filho et al. (2004), which indicate that in Rio Grande do Sul, the highest averages of solar radiation for 10-day-periods occur during December and January. Regarding the total rainfall, 663.4 mm were computed during the soybean cycle, with two moments of rain shortage of supply.

As showed in Figure 2, it was observed that in the three first ten-day periods, the total amount of water in the 2003/04 agricultural year was above the climatological normal. In November, the third ten-day period, and in December the first ten-day period, rainfall amounted around 60 mm and the second ten-day period in December was close to 140 mm, exceeding the normal rainfall value (100 mm) observed in time series. In December the third ten-day period and in January the first ten-day period showed less rain supply, making it the first moment of water scarcity.

According to Table 1 values, when the crop was at early flowering (R$_1$), the air temperature was higher but remained below 35° C and the minimum was above 10°C. At the end of the soybean cycle (cv. Fepagro RS-10), there were 1945.1 accumulated degree days   (ADD). Meteorological conditions at the experimental site for main phenological stages of soybean crop (Table 1), according to Fehr & Caviness (1977).

In some crops, changes in phenological stages depend basically on temperature. For soybean crops, high air temperature during the growth stage reduce time for the flowering phase (Major et al., 1975). Under tropical and subtropical conditions, low temperatures limit severely plant growth, that have photosynthetic capacity reduced due to the drop of quantum efficiency of Photsystem II as well as reduced activities of Photosystem I. There is also decrease in cycles of synthesis of stromal enzymes in C$_3$ plants (Allen & Ort, 2001).

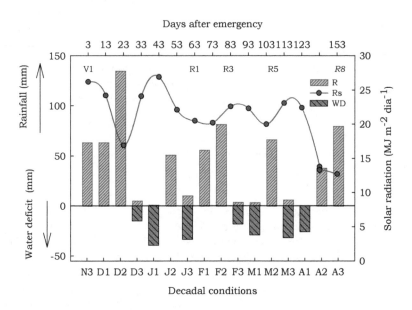

Fig. 2. Rainfall (R), solar radiation (Rs) and water deficit (WD) for ten-day periods during the soybeans cycle in the 2003/04crop year, in Eldorado do Sul, Brazil.

Growth stages	Rs	R	Air temperature (°C) (Average)			ADD	WD (mm)	
	MJ m⁻² dia⁻¹	(mm)	Tmax	Tmin	Tave	Tb (10°C)	CTNI	NTNI
S - V_{11}	22.6	328.5	28.5	16.4	22.5	789.6	101.2	93.7
V_{11} - R_1	20.1	0.0	34.3	19.4	26.9	806.5	3.1	3.5
R_1 - R_5*	21.8	141.2	29.4	16.5	22.9	1259.6	102.1	113.4
R_5** - R_7	19.2	76.1	29.3	17.1	23.2	1814.4	80.2	72.3
S - R_8	**20.6**	**663.4**	**28.6**	**16.3**	**22.5**	**1945.1**	**287.5**	**280.9**

Rs (Global solar radiation), R (rainfall), Tmax, Tmin, Tave (maximum, minimum and average air temperature), WD (water deficit). S (sowing), R_1 - beginning of flowering, R_5*- first day on R_5 period (beginning of grain filling); and R_5** - period between the second day on R_5 and the first day on R_7 (physiological maturity) and R_8 (complete maturity), ADD (accumulated degree-days calculated in Celsius unit, from seedling emergence on Nov.27.2003 to the end of cycle on Apr.30.2004). Tb is minimum base temperature for soybeans, 10°C.

Table 1. Meteorological parameters during the soybean cycle (cv. Fepagro RS-10) in different growth stages (Fehr & Caviness, 1977), in non-irrigated conventional tillage (CTNI) and no-tillage (NTNI). EEA / UFRGS, Eldorado do Sul, Brazil, 2003/04.

The results obtained in the field experiments showed that maximum evapotranspiration (ETm) amount during the cycle of soybeans was 681.3 mm, with a daily average of 4.5 mm day^{-1}. In the period between the late vegetative stage (V_{11}) and early grain filling (R_5), the highest evapotranspiration rates (6 to 8.0 mm day^{-1}) were observed between 49 and 93 DAE, with degree-days ranging between 789.6 and 1259.6. These rates reflect the condition of the irrigated crop in conventional tillage (lysimeter area) and maximum leaf area index (6.3), obtained in fully flowering (R_2).

It was observed in Martorano et al (2009) that on Dec 23.2003, the soil under conventional tillage systems without irrigation (CTI) contained 0.236 cm^3 cm^{-3} of moisture in the superficial layer (0.075m), indicating the beginning of drying, which was only observed in no-tillage irrigated (NTI) on Dec.25.2003, when soil water content was 0.270 cm^3 cm^{-3}. These data indicated the anticipation of the soil drying under conventional system compared to no-tillage. On Dec.26.2003, the soil was 0.185 cm^3 cm^{-3} in CTI and there was disruption of mercury columns of tensiometers on the most superficial layer (0.075 m), which was only observed in no-tillage on Dec.29, 2003, reinforcing the evidence of higher water storage in the upper layers in no-tillage systems.

In the subsequent depth of 0.15 m, the tensiometers Hg-columns disrupted in plots under conventional system on Dec.28.2003, while in no-tillage this fact occurred only on Jan.03.2004, showing that there is anticipation in soil drying front under conventional systems compared to no-tillage (Fig.3). By analyzing all the soil profile, observed that there was more rapidly advancing drying front in conventional system compared to no-tillage. In this first period of drying, the crop was in its growing period, and monitoring of plant responses to water conditions was focused on the second moment of less rainfall.

The second period of water shortage was from February, coinciding with the flowering and grain filling (from R_1 to R_5 stages), considered as critical in terms of water requirement for soybean crops (Berlato & Fontana, 1999). In this aspect, Martorano et al. (2009) showed the pattern of drying soil, indicating that the matric potential in the conventional system were more negative in relation to the no-tillage system. From 0.45 to 0.90 m of soil depth in the no-tillage and from 0.30 to 1.05 m in the conventional system, the soil remained drier than in the rest of the profile, probably due to a greater concentration of roots in these soil layers than in the surface layer, resulting from the tap root system of soybeans (Fig. 4). At 0.30 m deep, the matric potential in no-tillage was higher than -0.03 MPa, which was less negative than in the conventional system, around -0.05 MPa (Fig.5).

It is known that less soil moisture may place restrictions on water transfer to the atmosphere, not only influenced by the weather, but also by factors such as the root system, cultivar, management system, phytosanitary conditions and soil characteristics. For the conditions of the experiment, the main limiting factor was the reduction in water supply by rainfall, leading to periods of water deficit. The irrigated water treatment presents a brief discussion of the dynamics of soil water by simulated CROPGRO-Soybean and observed in field experiment showing the performance of the tool to simulate soil drying times. Fig. 6 shows that the moisture in the soil between 0.05 and 0.15 cm depths had differences management systems, indicating that no-tillage values showed higher humidity compared to conventional tillage irrigated condition, confirming the studies of Dalmago (2004) on the dynamics of drying the soil in these two management systems.

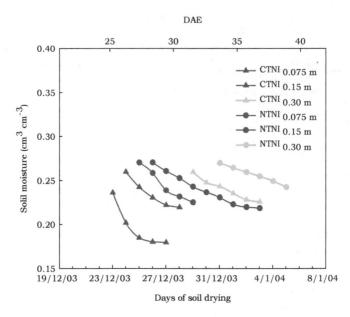

Fig. 3. Soil moisture in functions of days of soil drying and days after soybean plants emergence (DAE) under non-irrigated conventional tillage (CTNI) and no-tillage system (NTNI), in depths between 0.075 m and 0.30 m. Eldorado do Sul, Brazil, 2003/04, Brazil.

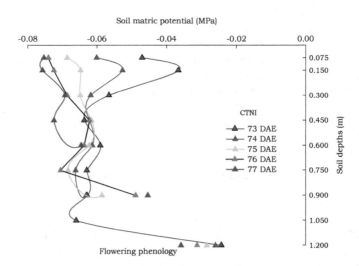

Fig. 4. Soil water matric potential in soybean crops under non-irrigated conventional system (CTNI) from 0.075 m to 0.30 m depths, at different days of soil drying, in Eldorado do Sul, Brazil, 2003/04.

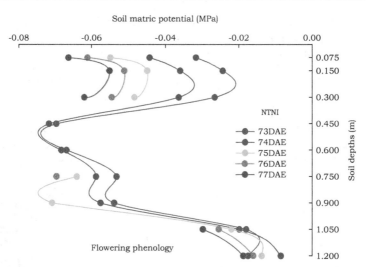

Fig. 5. Soil water matric potential in soybean crops under non-irrigated no-tillage system (NTNI), from 0.075 m to 0.30 m depths, at different days after plants emergence, in Eldorado do Sul, Brazil, 2003/04.

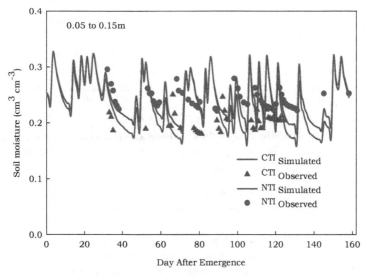

Fig. 6. Observed and simulated volumetric soil moisture at depths of 0.05 m and 0.15 m, in soybean crops under irrigated conventional tillage (CTI) and irrigated no-tillage (NIT) systems. Eldorado do Sul, Brazil, 2003/04.

In conventional tillage non-irrigated the model's ability was very low (32%), and no-tillage non-irrigated about 27% the value for the square root of the mean error, reinforcing the evidence of need for adjustments in routines or subroutines in the model to improve predictions of soil moisture levels.

According to Martorano et al (2007), the CROPGRO-Soybean was highly efficient in simulating the storage of soil water between 0.15 m and 0.30 m in conventional tillage and non-tillage irrigated. It was observed that in irrigated conventional tillage the model had high performance, with minimum distance between the simulated and observed (Fig. 7). The RMSE was 22.4% and in no-tillage irrigated (NTI) it was about 11%, which was considered a good precision of the model simulating water conditions in both systems adopted by farmers, showing that the tools can be use in management strategies in crops. The D-index of agreement (Willmott et al., 1985) in irrigated non-tillage (NTI) was 0.95, while the irrigated conventional tillage (CTI) had also a high level of agreement, 0.92, indicating high agreement closer to the 1:1 line (Fig. 7).

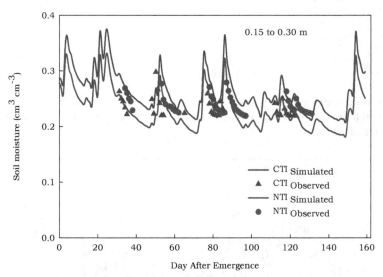

Fig. 7. Simulated soil moisture by CROPGRO-Soybean and observed in the layer between 0.15 and 0.30 m depths, in irrigated (I) plots of conventional tillage (CT) and no-tillage system (NT). Eldorado do Sul, Brazil, 2003/04.

In the same depth of non-irrigated treatments, the error (RMSE) in CTNI was 24.4% (which, depending on the boundary condition, may be acceptable) and NTNI was 19.0% in that limit, indicating good precision and accuracy of model, which can be used to simulate in this layer (Fig. 8). In irrigated conventional tillage the simulated values for the layers of 0.30 and 0.45 m depth may be considered acceptable presenting a 19.2% error (RMSE). In irrigated no-tillage it was 22.1% which, depending on the boundary condition, may be acceptable (Fig. 9).

In assessing the performance of CROPGRO for soybeans, after calibration of parameters such as hydraulic conductivity and root system depth in 224 points in 16 ha, using performance data for three years in Iowa, in the United States of America, Paz et al (1998) found that water stress explained 69% of income at all points measured, indicating the importance of adjustments in the soil parameters in crop yield simulations. In this sense, to increase the performance of CROPGRO Soybean tillage it is necessary to adjust the

Decision Strategies for Soil Water Estimations in Soybean Crops Subjected to No-Tillage and
Conventional Systems, in Brazil

191

parameters that simulate the limits of water retention in soil, capable of simulating the water supplies observed in the field, which determine water reserves income grain, in periods of soil drying.

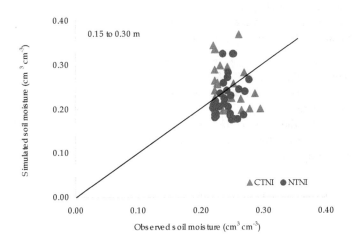

Fig. 8. Simulated soil moisture by CROPGRO Soybean and observed in a field experiment on non-irrigated conventional tillage (CTNI) and non-tillage (NTNI) systems, in the 0.15 to 0.30 m depths. Eldorado do Sul, Brazil, 2003/04.

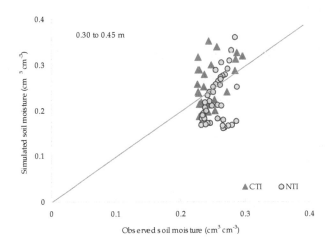

Fig. 9. Simulated soil moisture by CROPGRO Soybean and observed in a field experiment on irrigated conventional tillage (CTI) and non-tillage irrigated (NTI) systems in depths from 0.30 to 0.45 m. Eldorado do Sul, Brazil, 2003/04.

The model presented better performance in conventional tillage, simulating both the stock of water in the soil for growth and development and yield of soybean, than in the no-tillage system. The model simulations penalize indicators of growth and yield under no-tillage, regardless of the culture water status. The low and middle performance of the CROPGRO Soybean model for simulating the soil water inventories indicates that there is need for adjustments to the parameters that simulate the limits of water retention in soil, capable of simulating moisture observed in the field tillage.

Observing responses in the plant under no-tillage, in terms of growth and yield components, reinforce the effects of the water supply system during periods of dry soil. The comparison between observed and simulated data, through the CROPGRO-Soybean model, showed high accuracy of the simulated crop phenological stages, as demonstrated by the low scattering of points around the 1:1 line (Fig. 10), mostly for treatments with irrigation. These results may allow an efficient performance of the model in simulating the crop phenology (D-Index $\approx$ 1) and for estimating the canopy biomass.

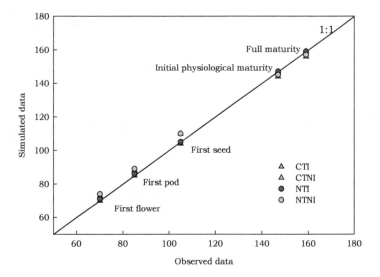

Fig. 10. Simulated by CROPGRO-Soybean model and observed phenological stages (in days after sowing) in CTI (irrigated conventional tillage), CTNI (non-irrigated conventional tillage), NTI (irrigated no-tillage) and NTNI (non-irrigated no-tillage) in Eldorado do Sul, Brazil, 2003/04.

Plant emergence occurred eight days after sowing (DAS) and the beginning of flower appearance was about 71 days after emergence, with a three-day difference between irrigated and non-irrigated treatments. In conventional tillage with irrigation, the comparison between simulated and observed data showed "D-Index" values corresponding to 0.83 for LAI, 0.96 for plant height, 0.93 for leaf weight, 0.97 for total dry biomass, 0.90 for pod weight and 0.98 for seed weight. In no-tillage system with irrigation, "D-Index" values were 0.82, 0.87, 0.89, 0.94, 0.88 and 0.91, respectively.

The model had lower accuracy under water deficit (non-irrigated treatments), especially in the no-tillage system. Regarding crop grain yields, the observed data were 3,597 kg ha^{-1} (irrigated conventional tillage), 3,816 kg ha^{-1} (irrigated no-tillage), 1,559 kg ha^{-1} (non-irrigated conventional tillage) and 1,894 kg ha^{-1} (non-irrigated no-tillage). The simulated grain yields by the model were 3,108 kg ha^{-1} (irrigated conventional tillage), 2,788 kg ha^{-1} (irrigated no-tillage), 824 kg ha^{-1} (non-irrigated conventional tillage) and 818 kg ha^{-1} (non-irrigated no-tillage).

5. Future research prospective

In order to increase efficiency of water used by plants, tools like DSSAT have been developed to help farmers with soil crop and water management planning. Researches on this tools applied to field experiments have shown that it is necessary to adjust models capable of simulating available water storage in the soil, especially in no-tilled cropping areas, for increasing the accuracy of simulations of the plant growth and grain yields of soybean by the CROPGRO-Soybean routines for soil water modeling (Faria & Madramootoo, 1996, Faria & Bowen, 2003) and soil C sequestration of DSSAT/CENTURY (Tornquist et al, 2009 a e b, Basso et al., 2011).

The models calibrated to deal with the soil system management have great potential to help developing ecologically efficient strategies related to water restitution in the soil, reducing the "Water Footprint" in irrigated tillage, increasing efficiency in food producing (FAO, 2006, FAO, 2009) and reducing agriculture water waste. Agrometeorological assessments help agricultural precision by indicating the right moment for water replacement during the crop phenological stages which are more vulnerable to water stress conditions.

On the other hand, impacts of climate changes related to water cycles, such as occurrence of extreme events and water supply for agriculture, represent major concerns for scientists and stakeholders dealing with environmental, social and economical analysis in different countries around the world. Particularly, the Brazilian economy is closely linked to the supply of natural resources, especially water use, both in agriculture and in hydropower generation, industrial sectors, and human consumption. The lack of adequate water management and planning can cause serious threats to human population. Although, there is plenty of drinking water in many regions of Brazil, this resource may become scarce if there is not any concern about the maintenance of water related ecosystem services. Major world problems occur due to impacts caused by human actions that disrespect the carrying capacity of natural environments.

Awareness of appropriate use by different productive sectors is indispensable to establish sustainable water management in economical activities. Water management should include monitoring and modeling schemes to assess the impact of economical activities on water sustainability, for instance, by using the "water footprint" approach to quantify agricultural water use for a particular product (Hoekstra & Chapagain, 2007). For agriculture, the concerns have turned toward water waste, and not only water scarcity in more arid regions of the world. Waste of water occurs due to improper decisions on water use, such as when, how and what is the most efficient technology to be applied in irrigated crops. Water deficit associated with periods of prolonged drought during the rainy season are a major cause of failed crops of grain in Brazil, especially in states in the Centre-South and Northeast.

Soil C sequestration (Lal, 2004) is reversible, as factors like soil disturbance, vegetation degradation, fire, erosion, nutrients shortage and water deficit may all lead to a rapid loss of soil organic carbon. It's the mechanism responsible for most of the mitigation potential in the agriculture sector, with an overall estimated 89% possible contribution to the technical potential (IPCC, 2007) excluding, however, the potential for fossil energy substitution through non agricultural use of biomass. There is a paucity of studies integrating soil C sequestration in the GHG balance of pastures and livestock systems (Soussana et al., 2007).

The Intergovernmental Panel on Climate Change (IPCC) pointed out that human actions are contributing to the increase of greenhouse gases (GHGs) in the atmosphere and have stepped up, especially extreme events with serious damage to humanity. Of all global anthropogenic CO_2 emissions, less than half accumulate in the atmosphere, where they contribute to global warming. The remainder is sequestered in oceans and terrestrial ecosystems such as forests, grasslands and peatlands (IPCC, 2007).

Considering concerns about climate changes, identification of soil, crop and livestock management to adapt and or mitigate GGE will be focused in new projects, for example, "AN Integration of Mitigation and Adaptation options for sustainable Livestock production under climate CHANGE – Animal Change" and "Role Of Biodiversity In climate change mitigation – ROBIN" in the Seventh framework program Food, Agriculture and Fisheries, Biotechnology (FP7). Further information and discussion about Brazilian researches related to livestock emissions of GHG are presented by Perondi et al (2011) and others in the site of "Global Research Alliance on Agricultural Greenhouse Gases (www.globalresearchalliance.org).

6. Conclusion

The soil water matric potential showed that drying front is longer in no-tillage system compared to conventional tillage and the CROPGRO-Soybean presented better performance to simulate phenological stages, growth variables and yield components under irrigated conditions than non-irrigated treatments, especially in conventional tillage. Crop yield results for no-tillage system presented low accuracy, mostly for water deficit, as shown by the test of Willmot, suggesting need for adjustment on model parameters to simulate soil water availability, especially for Brazilian agriculture, where no-tillage system area is increasing significantly.

7. Acknowledgement

The authors express their thanks to Brazilian Agricultural Research Corporation (Embrapa), the Federal University of Rio Grande do Sul (UFRGS), Institute Agronomic of

Parana (IAPAR), Foundation Research Agricultural of State Rio Grande do Sul (FEPAGRO) and the National Council for Scientific and Technological Development CNPq) for supporting scientific, financial and technological developments in this research. We would like also to thanks UFRGS/Agrometeorology Department to support scientific and friends for its support. The first author would like to thanks the colleagues Azeneth Schuler, Thomas Muello, Daiana Monteiro, Rodrigo Almeida, Leila Lisboa, Siglea Chaves and Salomão Rodrigues for having reviewed this paper.

8. References

Allen, D.J. & Ort, D.R. (2001). Impacts of chilling temperatures on photosynthesis in warm-climate plants. *Trends in Plant Science* Vol. 6, pp.36-42.

Bergamaschi, H., Guadagnin, M.R., Cardoso, L.S. & Silva, M.I.G. da. (2003). Clima da Estação Experimental da UFRGS (e Região de Abrangência). *Porto Alegre: UFRGS.* pp. 78.

Bergamaschi, H., Dalmago, A.G., Bergonci, J.I., Bianchi, C.A.M., Müller, A.G. & Comiran, F. (2004). Distribuição hídrica no período crítico do milho e produção de grãos. *Pesquisa Agropecuária Brasileira* Vol. 39, nº 9 (Set.), pp. 831-839.

Berlato, M.A. & Fontana, D.C. (1999). Variabilidade interanual da precipitação pluvial e rendimento da soja no Estado do Rio Grande do Sul. *Revista Brasileira de Agrometeorologia* Vol. 7, n.1, pp.119-125.

Basso, B., Gargiulo, O. Paustian, K., Robertson, G.P., Porter, C., Grace, P R. & Jones. J.W. (2011). Procedures for Initializing Soil Organic Carbon Pools in the DSSAT-CENTURY Model for Agricultural Systems. *Soil Science Society of America Journal* Vol. 75, Issue: pp. 1 - 69.

Buarque, S. C. (2006). *Construindo o Desenvolvimento Local Sustentável: metodologia de planejamento.* Ed. Garamond: São Paulo. Brasil. 177p.

Cargnelutti Filho, A., Matzenauer, R. & Trindade, J. K. da. (2004). Ajustes de funções de distribuição de probabilidade à radiação solar global no Estado do Rio Grande do Sul. *Pesquisa Agropecuária Brasileira* Vol.39, n.12, pp.1157-1166.

Dalmago, G.A. (2004). Dinâmica da água no solo em cultivos de milho sob plantio direto e preparo convencional. *Tese de Doutorado.* Porto Alegre: UFRGS, 2004. pp. 244.

Denardin, J.E., Kochhann, R.A., Flores, C.A., Ferreira, T.N. , Cassol, E.A., Mondrado, A. & Schwarezet, R.A. (2005). Manejo de enchurrada em Sistema Plantio Direto. Porto Alegre. pp. 88.

Dourado Neto, D., Nielsen, D. R., Hopmans, J. W. & Parlange, M. B. (2005). Soil water retention curve Vol. 1,0, Davis.

FAO (2006). Livestock's Long Shadow−Environmental Issues and Options. *Food and Agriculture Organization,* Rome, Italy.

FAO (2009). *How to feed the world in 2050.*

Faria, R. T. & Madramootoo, C. A., (1996). Simulation of soil moisture profiles for wheat in Brazil. *Agric. Water Manage.* 31, 35-49.

Faria, R. T. & Bowen, W. T. (2003). Evaluation of DSSAT soil-water balance module under cropped and bare soil conditions. *Brazilian Archives of Biology and Technology*, Curitiba Vol. 46, n. 4, pp.489-498.

Fehr, W.R. & Caviness, C.E. (1977). Stages of soybean development. *Ames: Iowa State University of Science and Technology*. pp. 11.

Global Research Alliance (2011). URL: http://www.globalresearchalliance.org/

Hoekstra, A.Y. & Chapagain, A.K. (2007). Water footprints of nations: Water use by people as a function of their consumption pattern. *Water Resources Management* Vol. 21, (Oct.), p. 35–48, 0920-4741.

Hoogenboom, G. (2000). Contribution of agrometeorology to the simulation of crop production and its applications. *Agricultural and Forest Meteorology* Vol. 103, p. 137-157.

Hoogenboom, G., Jones, J.W., Poter, C. H., Wilkens, P.W.; Boote, K.J., Batchelor, W. D.; Hunt, L. A. & Tsuji, G. Y. (2003). Decision Support System for Agrotechnology Transfer (Version 4.0) Vol. 1. *Honolulu: University of Hawaii*, pp.1-60.

International Consortium for Agricultural Systems Application. (2009). URL: http://www.icasa.net

IPCC (2007). Climate Change 2007. The Physical Science Basis. Contribution of Working Group I to the Fourth Assessment Report of the Intergovernmental Panel on Climate Change. *IPCC*, Cambridge University Press.

Jones, J.W., Hoogenboom, G., Porter, C.H., Boote, K.J., Batchelor, W.D., Hunt, L.A., Wilkens, P.W., Singh, U., Gijsman, A.J. & Ritchie, J.T., (2003) The DSSAT cropping system model. *European Journal Agronomy* Vol. 18 pp. 235-265.

Lal R. (2004). Soil carbon sequestration impacts on global climate change and food security. *Science 304*, pp. 1623–1627.

Loague, K. & Green, R.E. (1991). Statistical and graphical methods for evaluating solute transport models: overview and application. *Journal of Contaminant Hydrology* Vol. 7, pp.51-73.

Major, D.J.; Johnson, D.R. & Luedders, V.D. (1975). Evaluation of eleven thermal unit methods for predicting soybean development. *Crop Science* Vol. 15, p.172-174.

Martin, T. N.; Storck, L. & Dourado Neto, D. (2007). Simulação estocástica da radiação fotossinteticamente ativa e da temperatura do ar por diferentes métodos. *Pesq. Agropecuária Brasileira [online]*Vol. 42, n.9, pp. 1211-1219. ISSN 0100-204X. URL: http://dx.doi.org/10.1590/S0100-204X2007000900001.

Martorano, L.G. (2007). Padrões de resposta da soja a condições hídricas do sistema solo-planta-atmosfera, observados no campo e simulados no sistema de suporte à decisão DSSAT. 2007. pp.151 p. *Tese (Doutorado em Fitotecnia/Agrometeorologia)*, Universidade Federal do Rio Grande do Sul, Porto Alegre.

Martorano, L. G., Faria, R. T., Bergamaschi, H. & Dalmago, G. A. (2007). Avaliação do desempenho do modelo CROPGRO-Soja para simular a umidade em Argissolo do Rio Grande do Sul. In: *Congresso Brasileiro de Agrometeorologia*, 15, 2007. Anais Aracaju: SBA, Brasil, pp. 4.

Decision Strategies for Soil Water Estimations in Soybean Crops Subjected to No-Tillage and
Conventional Systems, in Brazil

197

Martorano, L. G., Faria, R. T. de; Bergamaschi, H. & Dalmago, G. A. (2008 a). Evaluation of the COPGRO/DSSAT model performance for simulating plant growth and grain yield of soybeans, subjected to no-tillage and conventional systems in the subtropical southern Brazil. *Rivista di Agronomia: an international Journal of Agroecosystem Management*, vol. 3, n° 3 supp. (Jul./Sept.), pp. 795-796, 1125-4718

Martorano, L.G., Bergamaschi, H., Faria, R.T. de., Dalmago, G.A., Mielniczuk, J.Heckler, B. & Comiran, F. (2008 b). Simulações no CROPGRO/DSSAT do Carbono no Solo Cultivado com Soja sob Plantio Direto e Preparo Convencional, em Clima Subtropical do Sul do Brasil. *Anais da XVII Reunião de Manejo e Conservação de Solo e Água*, Rio de Janeiro, Brasil.

Martorano, L.G, Bergamaschi, H., Dalmago, G.A., Faria, R.T. de., Mielniczuk, J. & Comiran, F. (2009). Indicadores da condição hídrica do solo com soja em plantio direto e preparo convencional. *Revista Brasileira de Engenharia Ambiental* Vol. 13, n° 4 (Out.), p. 397-405, 1415-4366.

Matzenauer, R., Barni, N.A. & Maluf, J.R.T. (2003). Estimativa do consumo relativo de água para a cultura da soja no Estado do Rio Grande do Sul. *Ciência Rural* Vol. 33 (Nov./Dez.), n°. 6, pp. 1013-1019, 0103-8478.

Perondi, P. A. O., Pedroso, A. de F., Almeida, R. G. de., Furlan, S., Barioni, L.G., Berndt, A., Oliveira, P. A., Higarashi, M., Moraes, S., Martorano, L., Pereira, L.G.R., Visoli, M., Fasiabem, M. do C. R. & Fernandes, A. H. B. M. (2011). Emissão de gases nas atividades pecuárias. In: Simposio Internacional sobre gerenciamento de resíduos agropecuários e agroindustriais. *Foz do Iguaçu*, Paraná, Brasil. pp. 69-73.

Paz, J. O., Batchelor, W. D., Colvin, T. S., Logsdon, S. D., Kaspar, T. C. & Karlen, D. L. (1998). Analysis of water stress effects causing spatial yield variability in soybeans. *Transactions of the American Society of Agricultural Engineers* Vol. 41, n.5, pp. 1527-1534.

Ruiz-Nogucira, B., Boote, K.J.& Sal, F. (2001). Calibration and use of CROPGRO-Soybean model for improving soybean management under rainfed conditions. *Agricultural Systems* Vol. 68, Oxon, p. 151-163.

Soussana, J.F.& Luescher A. (2007). Temperate grasslands and global atmospheric change: a review. *Grass and Forage Science* Vol. 62, 127-134.

Tornquist, C. G., Giasson, E. ; Mielniczuk, J. ; Cerri, C. E. P. & Bernoux, M. (2009a). Soil Organic Carbon Stocks of Rio Grande do Sul, Brazil. *Soil Science Society of America Journal* Vol. 73, p. 975-982.

Tornquist, C. G.; Gassman, P.W.; Mielniczuk, J., Giasson, E. & Campbell,T. (2009b). Spatially explicit simulations of soil C dynamics in Southern Brazil: Integrating century and GIS with i_entury. *Geoderma (Amsterdam)* Vol. 150, p. 404-414.

Van Genuchten, M. T. (1980). A closed form equation form prediction the hydraulic conductivity of unsatured soils. *Soil Science Society of America Journal* Vol. 44: 892-898.

Willmott, C.J., Akleson, G.S., Davis, R.E., Feddema, J.J., Klink, K.M., Legates, D.R., Odonnell, J. & Rowe, C.M. (1985). Statistics for the evaluation and comparison of models. Journal of Geophysical Research Vol. 90, n° C5 (Sept. 1985), pp. 8995-9005, 0958-305X.

Permissions

The contributors of this book come from diverse backgrounds, making this book a truly international effort. This book will bring forth new frontiers with its revolutionizing research information and detailed analysis of the nascent developments around the world.

We would like to thank Dr. Manish Kumar, for lending his expertise to make the book truly unique. He has played a crucial role in the development of this book. Without his invaluable contribution this book wouldn't have been possible. He has made vital efforts to compile up to date information on the varied aspects of this subject to make this book a valuable addition to the collection of many professionals and students.

This book was conceptualized with the vision of imparting up-to-date information and advanced data in this field. To ensure the same, a matchless editorial board was set up. Every individual on the board went through rigorous rounds of assessment to prove their worth. After which they invested a large part of their time researching and compiling the most relevant data for our readers. Conferences and sessions were held from time to time between the editorial board and the contributing authors to present the data in the most comprehensible form. The editorial team has worked tirelessly to provide valuable and valid information to help people across the globe.

Every chapter published in this book has been scrutinized by our experts. Their significance has been extensively debated. The topics covered herein carry significant findings which will fuel the growth of the discipline. They may even be implemented as practical applications or may be referred to as a beginning point for another development. Chapters in this book were first published by InTech; hereby published with permission under the Creative Commons Attribution License or equivalent.

The editorial board has been involved in producing this book since its inception. They have spent rigorous hours researching and exploring the diverse topics which have resulted in the successful publishing of this book. They have passed on their knowledge of decades through this book. To expedite this challenging task, the publisher supported the team at every step. A small team of assistant editors was also appointed to further simplify the editing procedure and attain best results for the readers.

Our editorial team has been hand-picked from every corner of the world. Their multi-ethnicity adds dynamic inputs to the discussions which result in innovative outcomes. These outcomes are then further discussed with the researchers and contributors who give their valuable feedback and opinion regarding the same. The feedback is then collaborated with the researches and they are edited in a comprehensive manner to aid the understanding of the subject.

Apart from the editorial board, the designing team has also invested a significant amount of their time in understanding the subject and creating the most relevant covers. They scrutinized every image to scout for the most suitable representation of the subject and create an appropriate cover for the book.

The publishing team has been involved in this book since its early stages. They were actively engaged in every process, be it collecting the data, connecting with the contributors or procuring relevant information. The team has been an ardent support to the editorial, designing and production team. Their endless efforts to recruit the best for this project, has resulted in the accomplishment of this book. They are a veteran in the field of academics and their pool of knowledge is as vast as their experience in printing. Their expertise and guidance has proved useful at every step. Their uncompromising quality standards have made this book an exceptional effort. Their encouragement from time to time has been an inspiration for everyone.

The publisher and the editorial board hope that this book will prove to be a valuable piece of knowledge for researchers, students, practitioners and scholars across the globe.

List of Contributors

Fernando Visconti
Desertification Research Centre – CIDE (CSIC, UVEG, GV), Valencia, Spain
Valencian Institute of Agricultural Research – IVIA Center for the Development of the Sustainable Agriculture - CDAS, Valencia, Spain

José Miguel de Paz
Valencian Institute of Agricultural Research – IVIA Center for the Development of the Sustainable Agriculture - CDAS, Valencia, Spain

Salwa Saidi, Salem Bouri and Hamed Ben Dhia
Water, Energy and Environment Laboratory (LR3E), ENIS, Sfax, Tunisia

Brice Anselme
PRODIG Laboratory, Sorbonne University, Paris, France

N. G. Shah
CTARA, IIT Bombay, India

Ipsita Das
Department of Electrical Engg, IIT Bombay, India

Lidija Tadić
Faculty of Civil Engineering, University of J. J. Strossmayer Osijek, Croatia

Francine Rochford
La Trobe University, Australia

Milan Cisty
Slovak University of Technology Bratislava, Slovak Republic

Mouhamadou Samsidy Goudiaby and Abdou Sene
LANI, UFR SAT, Université Gaston Berger, Saint-Louis, Senegal

Gunilla Kreiss
Division of Scientific Computing, Department of Information Technology, Uppsala University, Uppsala, Sweden

Wim Van Averbeke
Tshwane University of Technology, South Africa

Meijian Bai, Di Xu, Yinong Li and Shaohui Zhang
National Center of Efficient Irrigation Engineering and Technology Research, China Institute of Water Resources and Hydropower Research, China

Lucieta G. Martorano
Embrapa Amazônia Oriental Travessa Eneas Pinheiro s/n, Belém, PA, Brazil

Homero Bergamaschi
UFRGS and CNPq, P. Alegre, RS, Brazil

Rogério T. de Faria
FCAV-Unesp, Jaboticabal, SP, Brazil

Genei A. Dalmago
Embrapa Trigo, Passo Fundo, RS, Brazil

9 781632 391278